# ANALYTIC METHODS IN SPORTS

*Using Mathematics and Statistics to Understand Data from Baseball, Football, Basketball, and Other Sports*

# ANALYTIC METHODS IN SPORTS

*Using Mathematics and Statistics to Understand Data
from Baseball, Football, Basketball, and Other Sports*

## THOMAS A. SEVERINI

*Northwestern University
Evanston, Illinois, USA*

CRC Press
Taylor & Francis Group
Boca Raton   London   New York

CRC Press is an imprint of the
Taylor & Francis Group, an **Informa** business

A CHAPMAN & HALL BOOK

CRC Press
Taylor & Francis Group
6000 Broken Sound Parkway NW, Suite 300
Boca Raton, FL 33487-2742

© 2015 by Taylor & Francis Group, LLC
CRC Press is an imprint of Taylor & Francis Group, an Informa business

No claim to original U.S. Government works

ISBN-13: 9781482237016 (Hbk)

------------------------------------------------------------------------

**Library of Congress Cataloging-in-Publication Data**

------------------------------------------------------------------------

Severini, Thomas A. (Thomas Alan), 1959-
    Analytic methods in sports : using mathematics and statistics to understand data from baseball, football, basketball, and other sports / Thomas A. Severini.
        pages cm
    Includes bibliographical references and index.
    ISBN 978-1-4822-3701-6 (hardback : acid-free paper)
    1. Sports--Data processing. 2. Sports--Mathematical models. 3. Sports--Statistics. I. Title.

GV568.S48 2015
796.02'1--dc23                                                         2014006679

------------------------------------------------------------------------

**Visit the Taylor & Francis Web site at**
**http://www.taylorandfrancis.com**

**and the CRC Press Web site at**
**http://www.crcpress.com**

To Kate, Tony, Joe, and Lisa

# Contents

# Preface

One of the greatest changes in the sports world over the past 20 years or so has been the use of mathematical methods, together with the vast amounts of data now available, to analyze performances, recognize trends and patterns, and predict results. Starting with the "sabermetricians" analyzing baseball statistics, both amateur and professional statisticians have contributed to our knowledge of a wide range of sports.

There is no shortage of books, articles, and blogs that analyze data, develop new statistics, and provide insights on athletes and teams. Although there are exceptions (such as *The Book*, by Tango, Lichtman, and Dolphin [2007] and much of the work presented on FanGraphs at http//www.fangraphs.com), in many cases the analyses are ad hoc, developed by those with a knack for numbers and a love of sports. The fact that their methods have worked so well, and their results have been so useful, is a testament to the skill and ingenuity of the analysts.

However, just as formal methods of statistical analysis have contributed to, and improved, nearly every field of science and social science, these methods can also improve the analysis of sports data. Even when such methods are used in the context of sports, rarely is enough detail given for the beginner to understand the methodology. Although such methods are routinely covered in university courses on statistics, a student might need to take three or more courses before techniques that are useful in analyzing sports data are covered. Furthermore, the analysis of sports data has special features that generally are not covered in standard statistics courses and texts.

The goal of this book is to provide a concise but thorough introduction to the analytic and statistical methods that are useful in studying sports. It focuses on the application of the methods to sports data and the interpretation of the results; the book is designed for readers who are comfortable with mathematics but who have no previous background in statistics.

It is sometimes said that the key to an interesting and useful analysis in sports is asking the right questions. That may be true, but it is also important to know how to answer the questions. In this book, I try to give the reader the tools needed to answer questions of interest.

I would like to thank Karla Engel, who suggested the project and who helped in preparing the manuscript, including making many useful comments and corrections. I would also like to thank Rob Calver for several suggestions that greatly improved the book.

*Thomas A. Severini*
*severini@northwestern.edu*
*Evanston, Illinois*

# About the Author

**Thomas A. Severini** is currently professor of statistics at Northwestern University. His research areas include likelihood inference, nonparametric and semiparametric methods, and applications to econometrics. He is also the author of *Likelihood Methods in Statistics* and *Elements of Distribution Theory*. He received his PhD in statistics from the University of Chicago. He is a fellow of the American Statistical Association.

# List of Tables

# Introduction

## 1.1 ANALYTIC METHODS

Analytic methods use data to draw conclusions and make decisions. The challenge in using these methods is that the messages of the data are not always clear, and it is often necessary to filter out the noise to see the underlying relationships in the data. Therefore, one distinguishing feature of analytic methods is that they recognize the inherent randomness of data and they are designed to extract useful information in the presence of this randomness.

This is particularly important when analyzing sports data because we all know that the results of a game or other sporting event depend not only on the skill of the participants but also on "luck" and randomness, and separating the contribution of skill from that of luck is not always easy. A second difficulty in analyzing sports data is that a sporting event is a type of "observational study," a study in which important aspects are not under the control of the analyst; that is, we simply observe the data as they are generated, and in contrast to a controlled experiment, we cannot choose which players or teams participate in a particular event or in a given situation.

Given the emphasis on data analysis, it is not surprising that statistical concepts are central to methods presented in this book. Although statistical methodology is a vast topic, fortunately, there are a few central concepts and basic methods that can greatly improve our understanding of data and the processes that generated them. Knowledge in this area will be beneficial to all serious sports fans, whether they simply want to better understand the "new statistics" that have been proposed or whether they want to conduct their own statistical analyses.

There are at least two important roles of statistics in analytic methods. One is the use of statistical methodology designed to efficiently extract relevant information about measurements and their relationships. Statistical models are essential in this process. The models used to analyze sports data are generally empirical rather than based on some underlying theory. That is, the models used describe general features of relationships between variables; such models have been found useful in many fields and form the core of statistical methodology. What distinguishes statistical models from other types of models is the use of the concept of probability in describing relationships; such models are often described as "stochastic," as opposed to a "deterministic" model, in which probability does not play a role. However, statistical methods do more than recognize randomness; they filter it out, exposing the meaningful relationships in data.

When using statistical models of any type, it is important to keep in mind that they involve some idealization and simplification of complicated physical relationships. However, this does not diminish their usefulness; in fact, it is an essential part of it. Appropriate simplification is a crucial step in stripping away the randomness that often clouds our perception of the salient factors driving sports results.

A second role of statistical concepts in analytic methods is to give a framework for using probability to describe uncertainty. Given the random nature of the results of sporting events, any conclusions we draw from analyzing sports data will naturally have some uncertainty associated with them. Statistical methodology is designed to assess this uncertainty and express it in a useful and meaningful way. Explicit recognition of the random nature of sports is one of the primary contributions of analytic methods, such as those used in sabermetrics, to the analysis of sports data.

# 1.2 ORGANIZATION OF THE BOOK

Although a central theme of the book is the use of statistical models in understanding and interpreting sports data, before presenting the details of these methods, it is important to understand the basic properties of data. These properties are the subject of Chapter 2, which covers the fundamental methods of describing and summarizing data.

As noted in the previous section, the use of probability theory and statistical methodology to describe relationships and express conclusions is a crucial part of analytic methods. Chapter 3 covers those aspects of probability theory that are necessary to understand the randomness inherent in sports data. These concepts are applied to a number of scenarios in sports in which consideration of the underlying probabilities leads to useful insights. As noted previously, appreciating and understanding randomness is one of the main contributions of analytic methods.

Chapter 4 has several goals. One is to describe the statistical reasoning that underlies the analytic methods described in this book. Another is to present some basic statistical concepts, such as the *margin of error* and *statistical significance*, that play a central role in dealing with the randomness of sports data. Finally, Chapter 4 covers some basic statistical methods that are essential in studying sports data.

Chapters 5 through 7 develop the core statistical procedures for analyzing data based on sports results. Chapter 5 is concerned with detecting the presence of a relationship between variables and measuring the strength of such relationships. Several different methods are presented, designed to deal with different types of data and different goals for the analysis.

Chapter 6 takes the basic theme of Chapter 5—the relationship between variables—and goes a step further, covering methods for summarizing the relationship between two variables in a concise and useful way. These methods, known collectively as *linear regression,* use statistical methodology to find a function relating the two variables. The simplest method of this type yields a linear function for the variables; Chapter 6 also covers more sophisticated methods that are used when the relationship is nonlinear.

In Chapter 7, these methods are extended to the case of several variables when we wish to describe one of the variables, known as the *response variable*, in terms of the others, known as *predictors*. These methods, also known as linear regression, are perhaps the most commonly used statistical procedures, with applications in a wide range of scientific fields. Chapter 7 contains a detailed discussion of the basic methodology, along with more advanced topics such as the use of categorical variables as predictors, methods for finding the most important predictor, and interaction, which occurs when the effect of one of the predictors depends on the values of other predictors. In addition to the descriptions of the relevant statistical methodology, Chapters 6 and 7 include important information on the strengths and limitations of these methods as well as on the implementation of the methodology and the interpretation of the results.

The topics covered are similar to those that would be covered in courses on statistical methodology. However, they have been chosen specifically because of their importance and usefulness in analyzing sports data. Therefore, statistical methods that are not useful in analyzing sports data are not covered. Furthermore, many of the topics that are covered are fairly advanced in the sense that they would not typically be covered in an introductory statistics course.

# 1.3 DATA

The analytic methods described in this book have as their ultimate goal the analysis of data. Therefore, throughout the book, the methodology presented is illustrated on genuine data drawn from a wide variety of sports. Readers are encouraged to replicate these analyses, as well as to conduct related analyses using their own data.

There is no shortage of data available on the Internet. Some sites that have been found useful include sites operated by sports leagues or organizations (e.g., MLB.com, NFL.com, PGA.com, etc.); the sites operated by news organizations (e.g., ESPN.com, SI.com, yahoo.sports.com, etc.); the sports reference sites (e.g., Baseball-Reference. com, Pro-Football-reference.com); and independent sites such as FanGraphs.com. The "Reference.com" sites (e.g., Baseball-Reference.com and Pro-Football-Reference. com) are particularly noteworthy for the detailed data available and for their search engines, which are invaluable for finding data relevant to a specific question. Although some of these features require a modest subscription fee, for serious study of sports, the cost is minor.

For many of the examples given in this book, there are generally many possible sources for the data, and a specific source is not given. This is generally the case when the data analyzed are based directly on game results. Although all sites for a given sport contain the same basic data, different sites often contain different specialized data, such as "splits," data for specific situations, or "advanced statistics" that have been calculated from the basic data. Also, it is sometimes easier to retrieve or download data from some sites than others, depending on the user's software and experience. Therefore, it might be helpful to check a number of sources to find the ones that are most useful. When the data under consideration are more site specific, a reference to the source of the data is given.

Many of the datasets analyzed in this book are available in Excel format by following the link on my website (http://www.taseverini.com). When a dataset is available, it is denoted in the text by a label of the form Dataset C.x, where C denotes the relevant chapter and x denotes the dataset within that chapter; for example, the first available dataset in Chapter 2 is denoted as Dataset 2.1. A list of all available datasets, together with a description of the variables in the Excel spreadsheet and the source of the data, is given at the end of the book.

# 1.4 COMPUTATION

Nearly all the methods presented in this book require computation, and for the vast majority, some type of software is needed. One option is to use specialized statistical software, which is readily available. However, to make the techniques widely accessible, I assume that readers have only the spreadsheet program Excel available. Near the end of each chapter in which computation plays an important role, I have included a section outlining how to perform the calculations in Excel, enabling readers to repeat the analyses given in the book. These sections assume that the reader has basic knowledge of standard Excel functions; for others, there are many excellent books on Excel available. The commands listed in these sections are based on the 2010 version of Excel.

Statistical calculations in Excel require the Analysis ToolPak, which is available but is not automatically loaded when Excel is started. To load the Analysis ToolPak, use the following procedure, which is taken from the Excel help file.

1. Click the **File** tab, click **Options**, and then click the **Add-Ins** category.
2. In the **Manage** box, select **Excel Add-Ins** and then click **Go**.
3. In the **Add-Ins available** box, select the **Analysis ToolPak** check box and then click **OK**.

**Tip:** If **Analysis ToolPak** is not listed in the **Add-Ins available** box, click **Browse** to locate it. If you are prompted that the Analysis ToolPak is not currently installed on your computer, click **Yes** to install it.

The **Data Analysis** button will now be available under the **Data** tab.

For those readers who have access and experience with statistical software, such software may be used in place of Excel. In the present context, the main advantage of such programs is ease of use and presentation of results; statistical software also generally includes software for more sophisticated methods that are not used here.

# 1.5 SUGGESTIONS FOR FURTHER READING

Each chapter concludes with a section containing references and suggestions for further reading, as well as some occasional comments on the material covered in the main text.

Analysis of sports data is a rapidly growing field, and there are many books describing specific analyses. Here I give a few examples; there are many others. The work of Tango, Lichtman, and Dolphin (2007) contains a detailed treatment of how analytic methods can be used to answer specific questions about baseball strategy. Keri's (2006) work is a collection of interesting essays in which analytic methods are used to study issues ranging from a comparison of Babe Ruth to Barry Bonds (Silver, 2006a) to what baseball statistics can tell us about steroids (Silver, 2006c). Joyner (2008) provides similar analyses for football-related questions. The works of Moskowitz and Wertheim (2011) and Winston (2009) both show how analytic methods can be used to address a wide range of sports issues.

The mathematical and statistical topics discussed in this book are covered in many books on math and statistics; specific references are given in the relevant chapters. The work of Cox and Donnelly (2011, Chapter 6) contains a detailed discussion of the role of statistical models, expanding on the discussion in Section 1.1. Freedman (2009) discusses the theory and application of statistical models, paying particular attention to the analysis of observational data; much of that discussion is relevant to the analysis of sports data. Silver (2012) gives an entertaining, non-technical, introduction to analytic methods from the point of view of prediction; many of the points raised by Silver are useful in the analysis of sports data.

# Describing and Summarizing Sports Data

## 2.1 INTRODUCTION

Analytic methods use data, together with mathematical and statistical modeling, to gain information and make decisions. A vast amount of data is collected about sports, and this data collection can only be expected to grow in the future. The goal of this book is to describe methods that can be used to extract useful information from such data.

Although extracting information in this way generally involves sophisticated statistical techniques, before describing such methods it is important to have a good understanding of the basic properties of data. Such properties are the subject of this chapter.

In describing data, it is useful to think in terms of *subjects* and *variables*. A subject is an object on which data are collected; for sports data, the subjects are often players, but they might also be teams, games, seasons, or coaches. In some analyses, there is some subjectivity in how subjects are defined. For instance, suppose that we are studying the performance of NFL (National Football League) quarterbacks in the 2011 and 2012 seasons. We might treat each specific quarterback, such as Tom Brady, as a subject, with two sets of performance measures per quarterback, one for each year. On the other hand, we might treat the 2011 and 2012 versions of Brady as two different subjects. The choice depends on the goals of the analysis.

A variable is a characteristic of a subject that can be measured. For instance, if the subjects are MLB (Major League Baseball) players in the 2011 season, games played, at bats, hits, and assists are all examples of variables. The set of all variables for the subjects under consideration makes up the data to be analyzed. The appropriate type of analysis depends on the properties of these data.

## 2.2 TYPES OF DATA ENCOUNTERED IN SPORTS

An important property of a variable is the set of values it can take, called its *measurement scale*. A measurement scale is often a set of numbers. For instance, if our subjects are NFL running backs in the 2011 season and our variable is the number of carries in that season, then the measurement scale is the set of integers 0, 1, 2, ... . Note that, although there is a practical limit to the number of carries that a running back might have in a season, it is not usually necessary to specify that maximum in advance. A variable with a measurement scale consisting of numbers is said to be *quantitative*.

However, not all variables are numerical in nature. For instance, if our subjects are MLB players in the 2011 season and our variable is the way in which the player bats, then the set of all possible values is left (*L*), right (*R*), and switch hitter (*S*). Such a variable is said to be *qualitative* or *categorical*.

Although categorical variables are often important and useful, there are limitations to the type of statistical methods that can be used on such variables. For instance, in the batting example described, it does not make sense to refer to an "average batting style"; however, it would make sense to say that right-handed batters are most common or that 13% of players are switch hitters.

The measurement scale of a categorical variable might be ordered or unordered. For instance, in the batting example, there is no natural ordering of *R*, *L*, and *S*. Such a measurement scale is said to be *nominal*, and the variable is said to be a *nominal variable*. In other cases, the possible values of a categorical variable might have a natural ordering. For instance, suppose that the subject is a batter in baseball and the variable is the outcome of an at bat, with possible values single (*S*), double (*D*), triple (*T*), home run (*HR*), and no hit (*NH*), which includes outs as well as reaching base on an error or fielder's choice. Note that it is possible to divide *NH* into more specific outcomes (such as strikeout, fly out, etc.), resulting in a slightly different variable.

The five possible values *S, D, T, HR*, and *NH* are ordered, in the sense that *NH* is the least preferable, followed by *S, D, T*, and *HR* in order. Although this is a nominal variable, it has a little more structure than the variable measuring batting style. A nominal variable in which the possible values are ordered is said to be *ordinal*, and it is said to have an *ordinal scale*.

An ordinal variable can be converted to a quantitative variable by assigning numerical values to each possible value. For instance, in a total bases calculation, *NH* is assigned 0, *S* is assigned 1, *D* is assigned 2, *T* is assigned 3, and *HR* is assigned 4. Although there is some obvious logic to these choices, it is important to keep in mind that they are, to a large extent, arbitrary. Once such a numerical assignment is made, the resulting quantitative variable can be analyzed using the same methods used for other quantitative variables.

One way to think about the appropriateness of such a numerical assignment is to consider the distances between levels corresponding to the assignment. For instance, using the "total bases" variable, the difference between a single and a double is the same as the difference between a triple and a home run, and the difference between a single and a home run is the same as the difference between an out and a triple.

Of course, a different weighting system could be used, leading to a different quantitative variable. Consider the "isolated power" measurement, which is slugging average (often called slugging percentage) minus batting average. For a given player, let $N_S$, $N_D$, $N_T$, $N_{HR}$ denote the number of singles, doubles, triples, and home runs, respectively, and let $N$ denote the number of at bats. The formula for slugging percentage is

$$\frac{N_S + 2N_D + 3N_T + 4N_{HR}}{N}$$

and the formula for batting average is

$$\frac{N_S + N_D + N_T + N_{HR}}{N}.$$

It follows that the formula for isolated power is

$$\frac{N_S + 2N_D + 3N_T + 4N_{HR}}{N} - \frac{N_S + N_D + N_T + N_{HR}}{N} = \frac{N_D + 2N_T + 3N_{HR}}{N}.$$

Thus, isolated power uses an implicit weighting system of 0 for $NH$ and $S$, 1 for $D$, 2 for $T$, and 3 for $HR$.

Any weighting system that respects the ordering $NH \leq S \leq D \leq T \leq HR$ is valid. The challenge in converting an ordinal variable to a quantitative variable is to choose a weighting system that is meaningful for the goals of the analysis. For instance, both slugging average and isolated power are valid ways to summarize a player's offensive performance.

Quantitative variables are often classified as either *discrete* or *continuous*. Discrete variables are those whose set of possible values can be written as a list. For instance, suppose our subjects are NFL quarterbacks and we are analyzing performance in a given game. The variable "number of interceptions" takes values 0, 1, 2, ... and, hence, would be considered discrete.

A continuous variable is one that takes any value in a range. For instance, consider the quarterback example and the variable "completion percentage" measured for each game. Completion percentage takes any value between 0 and 100 and hence is a continuous variable.

All categorical variables are discrete. Quantitative variables might be either discrete or continuous. In some cases, a variable might be modeled as either discrete or continuous. Consider the quarterback example and the variable "completion percentage." A completion percentage of exactly 63.1% is technically impossible because it would require at least 1000 passing attempts; note that the fraction 631/1000 cannot be simplified. However, the concept of completion percentage is clearly a continuous one, and it would be treated as a continuous variable. That is, a completion percentage in a game of 63.1% makes sense, even if it is technically impossible, while 2.4 interceptions in a game does not.

some cases, a discrete variable might be treated as continuous if the number of values is relatively large. For instance, in the quarterback example, the variable "passing attempts" is technically discrete but would often be treated as continuous.

## 2.3 FREQUENCY DISTRIBUTIONS

The first step in analyzing a set of data is often some type of summarization. Consider the New York Yankees' 2011 season. If we are interested in the overall performance in the regular season, we could look at their game results, as presented in Table 2.1, where W denotes a win and L denotes a loss. The results are in the order that the games were played, across the rows, so that the first two games were wins, followed by a loss, for example. However, for many purposes this list contains too much detail; thus, we might summarize it by noting that in 2011 the Yankees won 97 games and lost 65 (see Table 2.2).

This is a simple example of a *frequency table,* and the information contained in the table is a *frequency distribution.* In statistical terminology, the number of wins is called the *frequency* of wins, and the percentage of wins is its *relative frequency.*

The frequency distribution of a categorical variable is defined in a similar manner: simply count the number of occurrences of each possible value of the variable to obtain the frequencies. The relative frequencies are found by dividing these counts by the total number of observations; relative frequencies can be reported as either proportions or percentages. If the variable is an ordinal one, the categories are generally placed in

**TABLE 2.1**  Wins and losses for the 2011 Yankees

| | | | | | | | | | | | | | | | | | | |
|---|---|---|---|---|---|---|---|---|---|---|---|---|---|---|---|---|---|---|
| W | W | L | W | L | W | L | W | L | W | W | L | W | W | L | W | W | W |
| L | L | W | W | L | W | W | W | L | L | L | W | L | W | W | L | L | L |
| L | L | L | W | W | W | L | W | W | L | W | W | L | L | W | W | W | W |
| L | W | W | L | L | L | W | W | W | L | W | W | W | L | W | W | W | W |
| L | L | W | W | W | W | W | W | W | L | L | W | L | L | W | W | L | L |
| W | W | W | L | W | L | W | L | W | W | W | L | L | W | W | W | W | W |
| W | W | W | L | L | L | W | W | L | W | W | W | L | W | W | L | W | L |
| L | W | L | L | W | W | W | L | W | W | W | W | W | W | L | L | L | L |
| W | W | W | L | L | W | L | W | W | W | W | L | W | W | L | L | L | L |

**TABLE 2.2**  Yankees Win–Loss Record in 2011

| OUTCOME | NUMBER | PERCENTAGE |
|---|---|---|
| Win | 97 | 59.9 |
| Loss | 65 | 40.1 |
| Total | 162 | |

**TABLE 2.3**  Braun's hits in 2011

| RESULT | COUNT | PERCENTAGE |
|--------|-------|------------|
| S | 110 | 58.8 |
| D | 38 | 20.3 |
| T | 6 | 3.2 |
| HR | 33 | 17.6 |
| Total | 187 | |

increasing order. A simple example is given in Table 2.3, which contains the results of Ryan Braun's 187 hits in 2011.

The frequency table for a discrete quantitative variable is handled in the same manner. For a continuous quantitative variable, the construction of a frequency table is a little more complicated because we cannot simply list the possible values of the variables. In these cases, we divide the range of the variable into nonoverlapping classes so that each observation falls into exactly one class. Then, the frequency and relative frequency of each class are determined as for a categorical variable. Table 2.4 contains a frequency table of Tom Brady's passing yards per game in games started from 2001 through the 2011 season (Dataset 2.1).

Therefore, there is some subjectivity in determining the frequency distribution of a continuous variable because of the subjectivity in how the classes are chosen. For example, in the Brady passing yards example, Table 2.5 gives another valid frequency table. The goal in choosing classes is to make the table useful for the analysis in mind.

In many cases, it is best to choose classes that are of equal length, as in the previous examples. However, in some cases it might be preferable to make some classes longer than others. One situation in which this occurs is when there are particular reference values of interest for the variable. For instance, in the Brady passing yards example, we might be interested in the classes less than 200, 200–299, 300–399, and at least 400 because these are values that are commonly used when discussing passing performance. The frequency table for these classes is given in Table 2.6.

Another situation in which classes of unequal length are often useful is when the variable analyzed is distributed unequally across its range. Consider the yards gained on Jamaal Charles's rushing attempts in 2010 (Dataset 2.2); these values are taken from

**TABLE 2.4**  Brady's passing yards in games started, 2001 to 2011

| CLASS | COUNT | PERCENTAGE |
|-------|-------|------------|
| 0–100 | 5 | 3.1 |
| 101–200 | 37 | 23.3 |
| 201–300 | 72 | 45.3 |
| 301–400 | 42 | 26.4 |
| 401–500 | 2 | 1.3 |
| 501–600 | 1 | 0.6 |
| Total | 159 | |

**TABLE 2.5**   Brady's passing yards in games started, 2001 to 2011

| CLASS | COUNT | PERCENTAGE |
|---|---|---|
| 0–50 | 1 | 0.6 |
| 51–100 | 4 | 2.5 |
| 101–150 | 13 | 8.2 |
| 151–200 | 24 | 15.1 |
| 201–250 | 39 | 24.5 |
| 251–300 | 33 | 20.8 |
| 301–350 | 23 | 14.5 |
| 351–400 | 19 | 11.9 |
| 401–450 | 2 | 1.3 |
| 451–500 | 0 | 0.0 |
| 501–550 | 1 | 0.6 |
| Total | 159 | |

the game play-by-play given on ESPN.com. The range of values is −6 to 80, but over 70% of the values are between 1 and 10 yards. A frequency table with equal-length classes, like the one in Table 2.7, either has many classes with relatively few observations or has very long classes. Although such a frequency table is still valid, it is difficult to choose the number of classes to summarize the data adequately without losing information. Table 2.8, which uses classes of varying length, seems to be better at summarizing the data.

Frequency tables are often presented graphically. Consider the data in Table 2.3, on Braun's hits in 2011. A graphical form of that table is given in Figure 2.1. This is just a bar chart of the class percentages; in statistics, such a table is called a *histogram*.

For continuous numerical data, the bars in the histogram are often shown to be touching to emphasize that the scale is continuous. For instance, a histogram for Brady's passing yards per game, based on the data in Table 2.5, is given in Figure 2.2.

If the classes are of unequal length, an adjustment is needed in constructing the histogram. The basic idea is that the area of the bars should be proportional to the frequency of the class. If the classes are of equal length, then this can be achieved by making the bar heights proportional to the frequency, but this does not work if the classes are of unequal length.

**TABLE 2.6**   Brady's passing yards in games started, 2001 to 2011

| CLASS | COUNT | PERCENTAGE |
|---|---|---|
| Less than 200 | 42 | 26.4 |
| 200–299 | 72 | 45.3 |
| 300–399 | 42 | 26.4 |
| 400 or more | 3 | 1.9 |
| Total | 159 | |

**TABLE 2.7**  Results of Charles's rushing attempts in 2010

| CLASS | COUNT | PERCENTAGE |
|---|---|---|
| −10 to −5 | 2 | 0.9 |
| −4 to 0 | 22 | 9.6 |
| 1–5 | 106 | 46.1 |
| 6–10 | 63 | 27.4 |
| 11–15 | 20 | 8.7 |
| 16–20 | 9 | 3.9 |
| 21–25 | 1 | 0.4 |
| 26–30 | 2 | 0.9 |
| 31–35 | 1 | 0.4 |
| 36–40 | 1 | 0.4 |
| 41–45 | 0 | 0.0 |
| 46–50 | 1 | 0.4 |
| 51–55 | 0 | 0.0 |
| 56–60 | 1 | 0.4 |
| 61–65 | 0 | 0.0 |
| 66–70 | 0 | 0.0 |
| 71–75 | 0 | 0.0 |
| 76–80 | 1 | 0.4 |
| Total | 230 | |

If we are interested in the values of frequencies or relative frequencies, then the frequency table is more useful than a histogram because these values are clearly presented. However, the histogram is useful for conveying the general pattern of frequencies, often referred to as the *shape* of the distribution.

The shape can be thought of as the information contained in the histogram if the $X$ and $Y$ scales are not given. In many respects, the "ideal" shape for a distribution is the familiar bell-shaped curve of the normal distribution; an example of such a histogram

**TABLE 2.8**  Results of Charles's rushing attempts in 2010

| CLASS | COUNT | PERCENTAGE |
|---|---|---|
| −1 or less | 16 | 7.0 |
| 0–2 | 51 | 22.2 |
| 3–5 | 63 | 27.4 |
| 6–10 | 63 | 27.4 |
| 11 -15 | 20 | 8.7 |
| 16–30 | 12 | 5.2 |
| 31–49 | 3 | 1.3 |
| 50 or more | 2 | 0.9 |
| Total | 230 | |

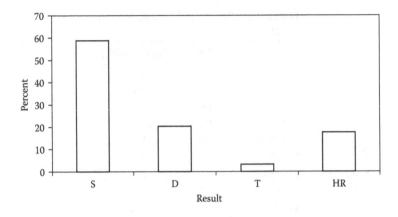

**FIGURE 2.1**   Histogram of Braun's hits in 2011.

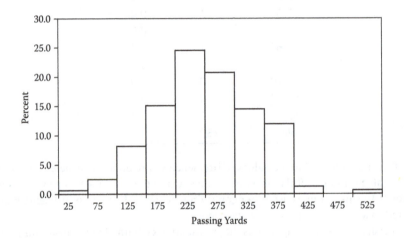

**FIGURE 2.2**   Brady's passing yards by game for 2001–2011.

**FIGURE 2.3**   Shape of the normal distribution.

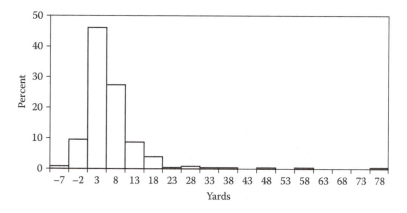

**FIGURE 2.4**  Charles's 2010 rushing yards by attempt.

is given in Figure 2.3. Of course, it is unrealistic to expect that any set of data will have a distribution that is exactly normal. However, the normal distribution can be used as a guide when assessing the distribution of a variable.

An important aspect of shape is *symmetry*; note that the normal distribution is exactly symmetric about its peak. The histogram in Figure 2.2, based on Brady's passing data, shows that these data are roughly symmetrically distributed with the peak of the distribution at about 250 yards with roughly the same number of observations near 300 yards as there is near 200 yards.

The shape of this histogram can be contrasted with the one based on Charles's rushing attempts data, given in Figure 2.4, based on the frequency table in Table 2.7. In this histogram, the peak of the distribution is at about 4 or 5 yards; however, there are several values much greater than 5 yards, and the minimum yardage on a carry is −6. Such a distribution is said to be *skewed*, and this property can be important in analyzing the data.

Another important property of the normal distribution is that it has only one peak or *mode*; such a distribution is said to be *unimodal*. The histograms in Figures 2.2 and 2.4 are both unimodal. A *bimodal* distribution has two peaks, separated by a "valley"; note that the two peaks do not have to be the same height. More generally, a distribution might have several modes.

For example, consider data on the shooting percentages of NBA (National Basketball Association) players for the 2010–2011 season; only "qualifying players," those with at least 300 field goals, are included (Dataset 2.3). A histogram of the shooting percentages is given in Figure 2.5. Note that this histogram has a peak around 44% and a second, smaller, peak at around 50%. A bimodal distribution often occurs when two subgroups, with different distributions for the variable under consideration, are combined.

Because the presence of a mode is sensitive to small changes in the data and to the exact intervals used in constructing a histogram, it is often difficult to say with certainty if a distribution is unimodal or bimodal simply by examining a histogram. However, if a bimodal distribution is suspected, it is generally a good idea to investigate further to see if a bimodal distribution is a possibility.

In the NBA shooting percentage example, we might expect different distributions of shooting percentage for guards and for forwards, which we take to include centers

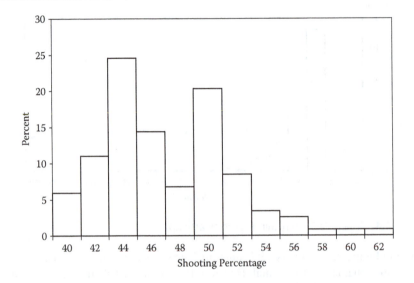

**FIGURE 2.5**   Shooting percentages of 2010–2011 NBA players.

as well. The players' positions are those assigned by ESPN.com. Figure 2.6 gives histograms of shooting percentages for these two subgroups. The distribution of shooting percentage for guards is unimodal and fairly symmetric.

However, the distribution for forwards does not have the classic unimodal shape. This is not surprising because there are many different styles of play for forwards; therefore, we could consider subdividing the forwards into small forwards and power forwards/centers. Figure 2.7 gives histograms of shooting percentages for these two groups.

Clearly, the distributions for small forwards and power forwards/centers are distinctly different. The distribution for power forwards/centers is still not exactly unimodal, although it is close. The lesson from the histograms is that any analysis of shooting percentages of NBA players will likely need to address the fact that shooting percentage is position dependent.

# 2.4 SUMMARIZING RESULTS BY A SINGLE NUMBER: MEAN AND MEDIAN

Although a frequency distribution provides some summarization of a set of data, in many cases a single-number summary is more useful. For quantitative data, the *mean* and *median* are the most commonly used summaries of this type. The mean of a set of data is simply the average, computed by finding the total of all observations and dividing by the number of observations. For example, for the data on Brady's passing yards in his 159 starts, the mean is 251.1 yards per game. Averages are commonly

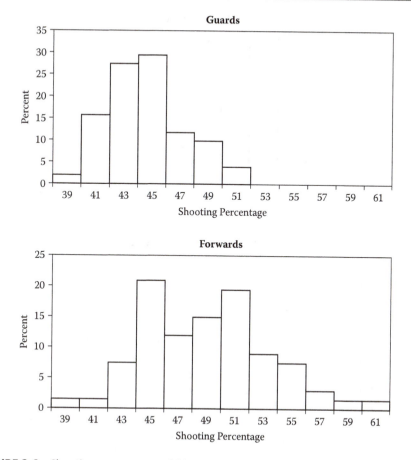

**FIGURE 2.6**    Shooting percentages of 2010–2011 NBA players by position.

used in traditional sports statistics. Some examples are points per game, yards per carry, and so on.

The *median* is often a useful alternative to the mean. The median is found by putting the data values in order and then finding the "middle value." Thus, the median has the interpretation that half the data values are above the median and half the data values are less than the median, roughly speaking. For the Brady passing yard data, the median value is 249 yards.

The mean and median are two different ways to summarize a set of data, and which is better in a given situation depends on the goals of the analysis. If the distribution is approximately symmetric, then the mean and median tend to be about the same. For example, we saw that the distribution of the Brady data is approximately symmetric, and the mean and median are close, 251.1 and 249, respectively. For a skewed distribution, however, the mean and median might be distinctly different because very large (or very small) observations have a greater effect on the mean than on the median, which depends primarily on the order of the data. In these cases, careful consideration should be given to the question of which measure to use.

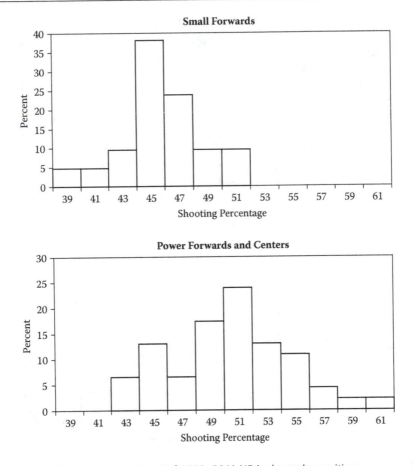

**FIGURE 2.7**   Shooting percentages of 2010–2011 NBA players by position.

For instance, consider the data on the yards gained on Charles's rushing attempts; the histogram in Figure 2.4 shows that this is a skewed distribution. The mean gain per attempt is 6.16 yards; the median is 4 yards. Therefore, about half the time, Charles gains 4 yards or less on a carry.

For most sports statistics, the mean and median are fairly close. Table 2.9 gives the mean and median for several offensive statistics for the set of 2011 MLB players with at least 502 plate appearances (Dataset 2.4). An exception is stolen bases, with a mean of 12.1 and a median of 8.0. This is not surprising because many players have few stolen bases but relatively few players have many stolen bases; that is, the distribution of stolen bases is highly skewed.

One difference between the mean and median is that the mean requires less-detailed data than the median. For instance, consider the yards gained on Charles's rushing attempts. We can calculate the average yards per carry using only his total yards gained and his total rushing attempts; calculation of the median requires the results of all 230 attempts. Another difference is that the mean is appropriate only for

**TABLE 2.9**  Offensive statistics for 2011 MLB qualifying players

| STATISTIC | MEAN | MEDIAN |
|---|---|---|
| Hits | 146.6 | 142.0 |
| Doubles | 29.1 | 27.0 |
| Triples | 3.2 | 2.0 |
| Home runs | 18.6 | 18.0 |
| Walks | 53.0 | 51.0 |
| Strikeouts | 101.7 | 97.0 |
| Stolen bases | 12.1 | 8.0 |
| Batting average | .272 | .269 |
| Slugging average | .442 | .446 |

quantitative data. Strictly speaking, the median can be calculated from ordinal data because we only need to be able to order the data values; that is, we do not need to be able to add data values in a meaningful way.

When analyzing ordinal data, often numerical values are assigned to the categories, as discussed in Section 2.2, and the mean can be calculated from the new, quantitative, variable. A simple example of this is batting average, in which any hit is assigned the value 1 and any nonhit is assigned the value 0. The batting average is simply the average of these values. More generally, any proportion can be viewed as an average of 0–1 variables.

A more complicated example is slugging average, in which the at-bat outcomes NH, S, D, T, and HR are assigned the values 0, 1, 2, 3, and 4, respectively; the slugging average is the average of these assigned values over a player's at bats.

It is important to keep in mind that any such weighting system is somewhat arbitrary, and the properties of the weighting system are reflected in the resulting mean value. One way to think about the appropriateness of such a numerical assignment for calculating averages is to consider sets of outcomes that yield equivalent averages under the assignment.

For example, consider two batters, each of whom has 100 hits. Suppose that the first batter has 150 singles, 10 doubles, and 10 triples; the second batter has 50 singles, 15 doubles, and 30 home runs. According to the total base calculation used in slugging average, each of these batters has 200 total bases; hence, they have the same slugging average. If this seems reasonable for the goals of the analysis, then the weighting system used in slugging average may be appropriate; if it does not, perhaps another weighting system would be preferable.

The median at-bat outcome can be calculated directly, without using any such weighting system. However, often the result is not particularly interesting. For instance, in 2011, Joey Votto had 599 at bats, which resulted in 113 singles, 40 doubles, 3 triples, 29 home runs, and 414 no hits. Therefore, his median result is no hit. In fact, this is true for any player with a batting average less than .500.

Rate statistics are closely related to averages. For example, consider runs allowed by 9 innings pitched for MLB pitchers. To calculate this, we can first compute runs allowed per inning by dividing runs allowed by innings pitched and then multiply this result by 9 to obtain runs allowed per 9 innings pitched. If pitchers always pitch complete innings, this is simply a rescaled version of the average value of a pitcher's runs allowed by inning. Because pitcher's can pitch a fraction of an inning, the run average is not exactly an average, but it is similar to one.

# 2.5 MEASURING THE VARIATION IN SPORTS DATA

Variation is an important part of all sports. The players in a league have varying skill sets, and their game performances vary. If two teams play each other several times, the outcomes will vary. Understanding, and accounting for, variation is a central goal of analytic methods. In this section some numerical measures of variability are presented; much of Chapters 4–7 is devoted to models used to describe and understand the variation that is inherent in all sports.

Consider the analysis of a quantitative variable. The variation of that variable refers to how the observed values of that variable differ from one another. One approach to measuring variation is to choose a reference value for the data and consider how the values differ from that reference value. A commonly used choice for the reference value is the mean of the data, although the median could also be used in this context.

The *standard deviation* (SD) of a set of data is, roughly speaking, the average distance from an observation to the mean of the data values. The formal definition of the standard deviation is a little more complicated—to compute the standard deviation, we first average the squared distances of the data values from the mean value; this is called the *variance* of the data. Although variances are sometimes used directly (see, e.g., Section 2.6), they have the drawback that the units are the square of the units of the observations themselves. For instance, if the measurement is in yards, the units of the variance will be yards squared. Therefore, we typically use the square root of the variance, which is called the standard deviation. The standard deviation may be viewed as the "typical" distance of a data value to the mean of the dataset. Note that the units of the standard deviation are the same as the units of the variable under consideration.

For example, consider the runs scored per game of the 30 MLB teams for the 2011 season. The standard deviation of these 30 values is 0.509; the standard deviation of the runs allowed per game of MLB teams for the 2011 season is 0.460. Table 2.10 contains the standard deviations of the points scored and points allowed per games of MLB, NFL, NBA, and NHL (National Hockey League) teams for the 2011, or 2010–2011, season. Note that for the NFL and MLB, there is more offensive than defensive variation in teams; for the NBA and NHL, the opposite is true.

One drawback of the standard deviation is that it is difficult to interpret directly. For instance, in Table 2.10, the standard deviation of points scored per game of NFL teams is more than 10 times greater than the standard deviation of runs scored of MLB teams, suggesting that NFL team offensive performance is much more variable than that of

**TABLE 2.10**   Standard deviations of teams' points scored and points allowed per game

| LEAGUE | STATISTIC | SD |
|--------|-----------|-----|
| MLB | Runs scored | 0.508 |
|  | Runs allowed | 0.458 |
| NFL | Points scored | 5.49 |
|  | Points allowed | 3.82 |
| NBA | Points scored | 4.04 |
|  | Points allowed | 4.88 |
| NHL | Goals scored | 0.268 |
|  | Goals allowed | 0.316 |

MLB teams. However, scoring in baseball is much lower than in football, in which most scoring plays are worth either 3 or 6 points. Therefore, we would expect the standard deviation of points scored for NFL teams to be greater than that of runs scored for MLB teams. However, it is difficult to say if the ratio of 10 is entirely caused by the difference in the way points are assigned in the two sports.

One reason for the difficulty in interpreting the standard deviation is that it is sensitive to the measurement units used; specifically, if data values are all multiplied by a constant $c$, then the standard deviation of the new data values is $c$ times the original standard deviation. For instance, suppose we measure NFL scoring in terms of "touchdown equivalents," points scored divided by 6; for example, if a team scores 33 points, we can express this as $33/6 = 5.5$ touchdown equivalents. Then, the standard deviation of the touchdown equivalents scored per game is

$$\frac{5.49}{6} = 0.915.$$

To eliminate the dependence on the units used, it is sometimes useful to express the standard deviation as a proportion, or percentage, of the average data value. The *coefficient of variation* (CV) is the standard deviation divided by the mean. Note that the coefficient of variation is a dimensionless quantity; therefore, it does not depend on the measurement scale: If all data values are multiplied by a constant $c$, both the standard deviation and the average of the new data values are multiplied by $c$ and $c$ cancels when taking the ratio. The coefficient of variation is generally only used for variables in which the values cannot be negative or 0.

Table 2.11 contains the coefficients of variation for points-scored and points-allowed variables analyzed in Table 2.10. Based on these results, NFL teams have the most variability in per game scoring, relative to the average points per game. Although the standard deviation of the points per game of NBA teams is relatively high compared to MLB, NFL, and NHL teams, as a percentage of the average points per game, it is the lowest of any of the four leagues.

Another approach to measuring the variation of a variable is to look at the spread of the values in the dataset. One such measure is the *range*, the maximum value minus

**TABLE 2.11**   Coefficients of variation of teams' points scored and points allowed per game

| LEAGUE | STATISTIC | CV (%) |
|--------|-----------|--------|
| MLB | Runs scored | 11.9 |
|  | Runs allowed | 10.7 |
| NFL | Points scored | 24.7 |
|  | Points allowed | 17.2 |
| NBA | Points scored | 4.1 |
|  | Points allowed | 4.9 |
| NHL | Goals scored | 9.6 |
|  | Goals allowed | 11.3 |

the minimum value. However, the range is too sensitive to occasional extreme values to be generally useful. An alternative approach is to base a measure of variation on the *quartiles* of the data. In the same way that the median is the "midpoint" of the set of data, the three quartiles divide the data into four parts, with roughly the same number of values in each part. Hence, the upper quartile is a value such that a fourth of the data lies above that value and three-fourths lie below; the lower quartile is a value such that a fourth of the data lies below that value and three-fourths lie above. The middle quartile is just the median.

The *interquartile range* (IQR) of the data is the upper quartile minus the lower quartile. Hence, the IQR gives the length of the interval containing the "middle half" of the data. The IQR offers at least two advantages over the standard deviation. One is that it has a more direct interpretation that is often useful in understanding the variability in a variable. Another is that it is less sensitive to extreme values than is the standard deviation, in the same sense that the median is less sensitive to extreme values than is the mean. Note, however, that, like the standard deviation, the IQR is sensitive to the measurement scale used, and multiplying each data value by a constant $c$ leads to the multiplication of the IQR by $c$.

Table 2.12 contains the standard deviation and IQR of the batting average (BA), on-base percentage (OBP), slugging average (SLG), and on-base plus slugging (OPS) for 2011 MLB players with a qualifying number of plate appearances. For instance, the

**TABLE 2.12**   Interquartile ranges and standard deviations of batting statistics

| STATISTIC | MEAN | MEDIAN | QUARTILES LOWER | UPPER | IQR | SD |
|-----------|------|--------|-------|-------|-----|-----|
| BA | .272 | .269 | .251 | .295 | .044 | .028 |
| OBP | .340 | .341 | .319 | .360 | .041 | .035 |
| SLG | .442 | .446 | .385 | .483 | .098 | .067 |
| OPS | .782 | .782 | .706 | .833 | .127 | .094 |

middle half of all batting averages falls in an interval of length 0.044, and half of all OPS values fall in an interval of length 0.127.

Note that, for all the variables listed, the IQR is larger than the standard deviation. This is not surprising because the IQR and standard deviation measure variation in different ways, and the relationship between IQR and standard deviation depends on the shape of the underlying distribution. For instance, for variables that follow a normal distribution, the standard deviation is approximately three-fourths of the IQR. For the variables in Table 2.12, the ratio of standard deviation to IQR ranges from 0.65 to 0.84. Note that, although variability in the data can be measured by the difference between the quartiles, it is often useful to present the quartiles themselves, as is done in Table 2.12.

# 2.6 SOURCES OF VARIATION: COMPARING BETWEEN-TEAM AND WITHIN-TEAM VARIATION

There can be many sources of the variation in a sequence of observations, and it is often interesting to consider the relative contributions of these different sources. This section considers this issue in the context of the variation in points (or runs) scored by teams over the course of a season.

Specifically, consider the variation in runs scored per game in the 2011 MLB season. Using data on all 162 games played for each of the 30 MLB teams, the standard deviation of runs scored is 3.01 runs. These data are taken from the game logs on Baseball-Reference.com.

Note that this variation in runs scored is caused by two factors: the variation in the run-scoring ability of the various teams, called the *between-team* variation, and the variation in each team's game-to-game run scoring, called the *within-team* scoring. Note that these two factors reflect different aspects of variation; between-team variation represents the differences between the 30 MLB teams, and the within-team variation represents the fact that the runs scored by a specific team vary considerably throughout the season.

Both types of variation can be measured by standard deviations, calculated from sets of observations appropriate to the variation under consideration. For instance, to measure between-team variation, we can use the standard deviation of the team-by-team average runs scored for the season; this value, 0.508, is the one given previously, in Table 2.10.

To measure the within-team variation, we can calculate the standard deviation of runs scored for each of the 30 teams and average the results. Table 2.13 contains the standard deviation for runs scored for 10 of the 30 teams. The average standard deviation for these 10 teams is 2.97 runs. If all 30 teams are analyzed, the average standard deviation is 2.98 runs.

Therefore, we have three measures of variation of runs scored in MLB games: the overall variation, the between-team variation, and the within-team variation. The

**TABLE 2.13** Averages and standard deviations of runs scored for selected teams

| TEAM | AVERAGE | SD |
| --- | --- | --- |
| Blue Jays | 4.59 | 3.19 |
| Cardinals | 4.70 | 2.92 |
| Giants | 3.52 | 2.71 |
| Mariners | 3.43 | 2.61 |
| Phillies | 4.40 | 3.02 |
| Pirates | 3.77 | 2.49 |
| Rockies | 4.53 | 3.39 |
| Tigers | 4.86 | 3.04 |
| White Sox | 4.04 | 2.69 |
| Yankees | 5.35 | 3.63 |

overall variation is measured by $S_O$, the standard deviation of the set of all 4860 runs-scored values in all games played in 2011; here, $S_0 = 3.01$. The between-team variation is measured by $S_B$, the standard deviation of the average runs scored for the 30 MLB teams; here, $S_B = 0.508$. The within-team variation is measured by $S_W$, obtained by computing the standard deviation of the runs scored for each team and then averaging these 30 values; here, $S_W = 2.98$.

Not surprisingly, these three measures are related. Specifically, $S_O^2$ is approximately equal to $S_B^2 + S_W^2$. This approximation is valid whenever both the number of teams under consideration and the number of observations for each team are relatively large, say greater than 10. Because here we have 30 teams and 162 observations per team, we expect the approximation to hold extremely well, and it is easy to see that it does. See Section 2.11 for references to more general results that can be applied in small-sample settings.

Partitioning the overall variation in this way allows us to consider the proportion of variation in runs scored per game that is due to the fact that teams have different offensive capabilities. Specifically, since $S_0^2 = 9.06$ and $S_B^2 = 0.26$, and

$$\frac{0.26}{9.06} = 0.029$$

we can conclude that approximately 3% of the variation in scoring in MLB games is caused by the variation between teams. The other 97% is because of the variation within each team, which can be attributed to a number of factors, including the differing abilities of opposing pitchers and the natural variation in run scoring in baseball.

This result can be contrasted with a similar computation for scoring in NFL games. Using results from the 2011 NFL season, $S_0 = 10.21$ and $S_B = 5.49$; it follows that approximately 29% of the variation in per game scoring in the NFL can be attributed to differences in team offensive abilities. One way to express this difference between MLB and NFL games is to say that the scoring in NFL games is much more predictable than is the scoring in MLB games; the analysis in this section allows us to quantify this statement.

# 2.7 MEASURING THE VARIATION IN A QUALITATIVE VARIABLE SUCH AS PITCH TYPE

The concept of variability is most commonly applied to quantitative variables. However, in some cases, we are interested in measuring the variability in qualitative variables. For instance, consider the types of pitches thrown by Clayton Kershaw during the 2012 baseball season. Using the PITCHf/x data as reported on FanGraphs.com, 62.5% of Kershaw's pitches were fastballs, 22.6% were sliders, 11.2% were curveballs, 3.4% were changeups, and 0.3% did not fall into one of the recorded categories and are labeled as "other." For comparison, consider the pitch distribution of Cole Hamels, who threw 51.3% fastballs, 8.9% curveballs, 30.3% changeups, 9.1% cutters, and 0.4% other. We might be interested in determining which of these pitchers had more variability in his pitch selection, or more generally, we might be interested in how variability in pitch type is related to success on the mound.

Note that the variable analyzed here, "pitch type," is qualitative; therefore, measures of variability based on the standard deviation do not apply. Instead, we focus on how the variable values are distributed across the various categories. Consider the following categories for pitches: fastball (FA), slider (SL), curveball (CU), changeup (CH), cutter (FC), sinker (SI), and other (O). For a given pitcher, let $p_{FA}$, $p_{SL}$, $p_{CU}$, $p_{CH}$, $p_{FC}$, $p_{SI}$, $p_O$ denote the proportions of pitches in each of the categories, respectively. For example, for Kershaw,

$$p_{FA} = 0.625, \quad p_{SL} = 0.226, \quad p_{CU} = 0.112, \quad p_{CH} = 0.034, \quad p_O = 0.003;$$

here, $p_{SI}$ and $p_{FC}$ are 0.

A measure of variability of pitch type is a function of these seven proportions. There are a few basic properties such a measure should satisfy: It should be nonnegative and equal to 0 only if all the pitches are of one type, and it should take its maximum value if each pitch type is equally likely because that is the most variation we can have in the pitch distribution.

One measure of variability satisfying these requirements is the *entropy*. The entropy is given by

$$-[p_{FA} \ln(p_{FA}) + p_{SL} \ln(p_{SL}) + p_{CU} \ln(p_{CU}) + p_{CH} \ln(p_{CH}) + p_{FC} \ln(p_{FC}) + p_{SI} \ln(p_{SI}) + p_O \ln(p_O)]$$

where *ln* denotes the natural log function, and $0\ln(0)$ is interpreted as 0. General properties of the natural log function are discussed in the following section. For calculating the entropy, we only need to be able to calculate natural logs; this can be done on most calculators as well as in Excel.

The entropy can be interpreted as the "predictability" of the pitch type based on the observed proportions; it is used in many areas of science. Because $\ln(1) = 0$, if one proportion is 1 while the others are all 0, the entropy is 0; otherwise, it is positive. It

**TABLE 2.14**    Top 5 and bottom 5 pitchers in pitch variability in 2012

| | STANDARD ENTROPY | | STANDARD ENTROPY |
|---|---|---|---|
| Joe Blanton | 0.909 | R. A. Dickey | 0.250 |
| Felix Hernandez | 0.901 | Ervin Santana | 0.481 |
| Barry Zito | 0.897 | Derek Holland | 0.496 |
| Paul Maholm | 0.895 | Jordan Zimmermann | 0.515 |
| C. J. Wilson | 0.883 | Clayton Kershaw | 0.518 |

may be shown that the maximum value of the entropy is achieved if all the proportions are equal; in the pitch-type example, this maximum value is $\ln(7)$. The standardized entropy can be calculated by dividing the entropy by this maximum value; the standardized entropy then lies in the interval from 0 to 1.

For Kershaw, the entropy of his pitch distribution is

$$-(0.625\ln(0.625) + 0.226\ln(0.226) + 0.112\ln(0.112) + 0.034\ln(0.034)$$
$$+ 0.003\ln(0.003)) = 1.0075;$$

Because $\ln(7) = 1.9459$, the standardized entropy of Kershaw's pitch distribution is 0.518. For comparison, the standardized entropy of Hamel's pitch distribution is 0.596; it follows that there is more variability in Hamel's pitch selections than there is in Kershaw's.

Table 2.14 contains the top and bottom five MLB pitchers in terms of pitch variability for the 2012 season. Only pitchers with at least 162 innings pitched were considered so that only starting pitchers are included. Knuckleballer R. A. Dickey had the least variability in his pitches, while Felix Hernandez, known for his large repertoire of pitches, had one of the highest values of standardized entropy.

# 2.8 USING TRANSFORMATIONS TO IMPROVE MEASURES OF TEAM AND PLAYER PERFORMANCE

Consider a measurement $X$ that we wish to analyze; for instance, $X$ could be an NFL quarterback's number of touchdown passes or an MLB player's home run rate, the number of home runs he hit in a given season, divided by his at bats (plate appearances could be used in place of at bats here). When working with such a measurement, we always have the option of transforming it by first applying a given function. That is, let $g$ be a function and let $Y = g(X)$. Then, instead of analyzing $X$, we could analyze $Y$. The function $g$ is known as a *transformation*.

Such transformations are generally chosen to be either strictly increasing or strictly decreasing, so that the order of the possible values of $X$ does not change. For instance, if $g$ is an increasing function, such as $g(x) = x^2$ (when applied to nonnegative $x$) and $X_1$ and $X_2$ are two possible values of $X$ such that $X_1 > X_2$, then $Y_1 = g(X_1)$ is greater than $Y_2 = g(X_2)$. If $g$ is a decreasing function, then $Y_1 < Y_2$. A transformation that is either increasing or decreasing is said to be *monotone*.

Because we can always recover the original $X$ from the transformed variable $Y$, $Y$ contains the same information as $X$. Therefore, the advantage of using the transformed variable as opposed to the original variable is to simplify the analysis and to aid in the interpretation of the results rather than to introduce new information.

Although, in principle, any monotone function $g$ can be used as a transformation, in practice there are only a few that are regularly used. These include the logarithm function $g(x) = \ln(x)$ and the inverse function $g(x) = 1/x$. Note that linear functions, which are of the form $g(x) = ax + b$ for some constants $a$ and $b$, are not used in this context. Linear functions correspond to a change in the units of the measurement, but they do not change other aspects of the data, such as the shape of the distribution and the relative spacing between values.

This is not to say that linear transformations cannot be useful. It is often helpful to recode a variable so that the range of values is more convenient. For instance, if the data are the interception rates of NFL quarterbacks (interceptions divided by passing attempts), it might be more convenient to use percentages instead of proportions so that Peyton Manning's 11 interceptions in 400 attempts in 2012 is recorded as 2.75 instead of as 0.0275, for example.

As noted, the logarithm or log is a commonly used transformation. Here, we use "natural logs," denoted by *ln*. It is important to note that $\ln(a)$ can only be calculated when $a$ is a positive number. The natural log of $a$, $\ln(a)$, is defined by the relationship

$$e^{\ln(a)} = a$$

where $e$ is a number known as the base of the natural logarithms; note that the function $e^x$ is sometimes written $\exp(x)$. The value of $e$ is approximately 2.71, but it is the properties of the logarithm that are implied by the defining relationship that are important, rather than the specific numerical values. Logarithms with other bases, such as base 10 or base 2 logarithms, are sometimes used; using a different base changes the numerical values of the logarithms, but it does not change their basic properties.

For instance, if $a$ and $b$ are positive numbers, then

$$\ln(ab) = \ln(a) + \ln(b)$$

so that the ln function transforms multiplication into addition in a certain sense. Because raising a number to an exponent is essentially multiplying the number by itself a certain number of times, we also have the property that

$$\ln(a^b) = b\ln(a).$$

The exponential function has related properties:

$$\exp(a+b) = \exp(a)\exp(b) \quad \text{and} \quad \exp(a)^b = \exp(ab).$$

Let $X$ be a given measurement and let $Y = \ln(X)$ or, equivalently, $X = \exp(Y)$. Consider a 1-unit increase in $Y$ from $Y$ to $Y + 1$ and let $X_1$ denote the $X$ value corresponding to $Y + 1$. Then,

$$X_1 = \exp(Y+1) = \exp(1)\exp(Y) \doteq 2.71\exp(Y) = 2.71\ X;$$

That is, increasing $Y$ by 1 unit corresponds to a proportional increase in $X$ from $X$ to $2.71X$. Therefore, a log transformation is often useful when it is believed that proportional differences in the variable are more relevant to the analysis than are additive differences. Such transformations are often used in science; for instance, the Richter scale, used to measure the energy released during an earthquake, and decibels, used to measure sound levels, are both based on log transformations.

For example, consider the earnings of golfers on the 2012 PGA (Professional Golfers' Association) tour (Dataset 2.5). Table 2.15 lists the top 10 money winners along with the 20th, 30th, 40th, 50th, and 100th top winners, as recorded on ESPN.com; it also gives each player's winnings in millions of dollars.

**TABLE 2.15**   Winnings in millions of dollars for PGA golfers in 2012

| RANK | PLAYER | WINNINGS |
|---|---|---|
| 1 | McIlroy | 8.05 |
| 2 | Woods | 6.13 |
| 3 | Snedeker | 4.99 |
| 4 | Dufner | 4.87 |
| 5 | Watson | 4.64 |
| 6 | Johnson | 4.50 |
| 7 | Rose | 4.29 |
| 8 | Mickelson | 4.20 |
| 9 | Mahan | 4.02 |
| 10 | Bradley | 3.91 |
| 20 | Garrigus | 3.21 |
| 30 | Curtis | 2.49 |
| 40 | Every | 1.97 |
| 50 | Byrd | 1.62 |
| 100 | Love | 0.99 |

If we compare players by simply taking the difference of their winnings, Johnson is equally close to McIlroy and Love because he made about $3.5 million less than McIlroy and about $3.5 million more than Love. However, we could also compare golfers by looking at the ratios of their winnings. Because Johnson made about 56% as much as McIlroy and Love made only about 22% as much as Johnson, this suggests that Johnson's 2012 performance was much closer to that of McIlroy than to that of Love.

An analysis based on ratios of winnings corresponds to comparing golfers in terms of their log-winnings. Therefore, in some sense, the choice between measuring winnings in raw dollars or using a log transformation comes down to whether considering Johnson's performance as halfway between those of McIlroy and Love or considering Johnson's performance as much closer to McIlroy's than to Love's seems to make more sense for analysis. Although the choice could vary depending on the goals of the analysis, in most cases it seems more appropriate to view Johnson's performance as closer to McIlroy's.

Table 2.16 contains the same players as in Table 2.15 but now with performance measured by log-winnings, with log-dollars as the units. Note that a difference of 1 unit in log-winnings corresponds to a ratio of about 2.7 in winnings measured in dollars (that is, not transformed). For instance, Johnson's log-winnings were 15.32, while Byrd's

**TABLE 2.16**  Winnings in log-dollars for PGA golfers in 2012

| RANK | PLAYER | LOG-WINNINGS |
|------|--------|--------------|
| 1 | McIlroy | 15.90 |
| 2 | Woods | 15.63 |
| 3 | Snedeker | 15.42 |
| 4 | Dufner | 15.40 |
| 5 | Watson | 15.35 |
| 6 | Johnson | 15.32 |
| 7 | Rose | 15.27 |
| 8 | Mickelson | 15.25 |
| 9 | Mahan | 15.21 |
| 10 | Bradley | 15.18 |
| 20 | Garrigus | 14.98 |
| 30 | Curtis | 14.73 |
| 40 | Every | 14.49 |
| 50 | Byrd | 14.30 |
| 100 | Love | 13.81 |

log-winnings were 14.30, a difference of about 1 unit. In terms of raw winnings, Johnson won 4.50 million, and Byrd won 1.62 million, a ratio of 4.50/1.62 = 2.8.

For these data, log-winnings appear to be a more useful measure of performance than the raw winnings values. However, there is an important drawback to using a log transformation. The transformed values are more difficult to interpret than are the original, untransformed, values. This is largely because of unfamiliarity, and if such a transformed measurement were routinely used, this drawback would be expected to be less important or even to disappear.

In many cases, it is possible to evaluate the suitability of a transformation by looking at the distribution of the transformed data values. Measurements for which the interpretation of one unit is the same throughout the range of the variable tend to have a distribution that is roughly symmetric and bell shaped. This occurs because many, if not most, of the underlying factors driving athletic performance are approximately normally distributed. Therefore, statistics that accurately measure some aspect of performance are often, although not always, approximately normally distributed as well.

Figure 2.8 contains histograms of the winnings of the 258 golfers listed in the PGA rankings, measured in millions of dollars, together with the histogram of log-winnings, measured in log-dollars. Note that each histogram is an accurate portrayal of the data used in the calculation. However, if we want to use the players' winnings as a measure of their performance in 2012, the two histograms tell vastly different stories.

The histogram of the raw winnings suggests that there was one outstanding golfer and a few very good golfers, and the vast majority of golfers performed relatively poorly compared to these few top golfers. The histogram of log-winnings suggests a distribution of performances that is more bell shaped, although still showing some asymmetry, with the majority having an "average" performance and a few performing either very well or very poorly. For most, but not all, purposes, a performance measure based on log-winnings appears to be more useful.

# 2.9  HOME RUNS PER AT BAT OR AT BATS PER HOME RUN?

Consider a rate statistic of the form $A/B$, where $A$ is a measure of achievement, such as yards gained or games won, and $B$ is a measure of participation, such as at bats or games played. For such a statistic, we could alternatively use $B/A$, and the choice between $A/B$ and $B/A$ as a measure of performance is essentially the same as choosing whether or not to transform a given statistic using the inverse transformation $1/x$.

In this section, we consider this issue for the case of measuring home run performance for MLB players. Therefore, let $X$ denote a player's home run rate, measured in home runs per at bat. An alternative measure to use when studying home run performance is $Y = 1/X$, at bats per home run. For example, in 2011 Jose Bautista hit 43 home runs in 513 at bats. This could be expressed as either 0.0838 home runs per at bat or as 11.9 at bats per home run.

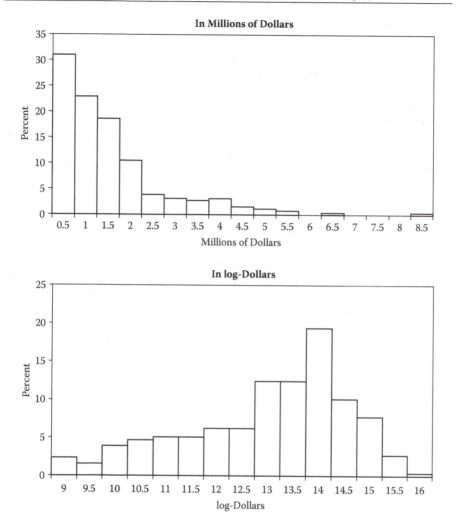

**FIGURE 2.8** The 2012 winnings of PGA golfers.

Although transforming measurements using a monotone transformation does not change the ordering of the values, it does change the spacing between them, as discussed in the previous section. For instance, in 2011, Mark Reynolds had 37 home runs in 534 at bats, which is 0.0693 home runs per at bat; Alfonso Soriano had 32 home runs in 585 at bats, which is 0.0547 home runs per at bat. Therefore, Reynolds's home run rate was about halfway between that of Bautista and Soriano because (0.0838+0.0547)/2 = 0.0693. Now, suppose we measure home run performance in terms of at bats per home run. Reynolds had 14.4 at bats per home run, and Soriano had 18.3 at bats per home run. In terms of this measure, Reynolds's performance was much closer to Bautista's than to Soriano's because 14.4 − 11.9 = 2.5 and 18.3 − 14.4 = 3.9.

One goal in choosing between home runs per at bat and at bats per home run is to have a measurement such that the importance of one "unit" of the measurement is the same throughout its entire range. For instance, in the present example, it is reasonable to evaluate home run performance in terms of how many home runs a player would hit in a "standard season" of, say, 600 at bats. If a player hits 0.02 home runs per at bat, we would expect $(0.02)(600) = 12$ home runs. For each increase of 0.01 in home run rate, we would expect 6 additional home runs.

Now, suppose we use at bats per home run to evaluate performance. If a player has 50 at bats per home run, this translates to $600/50 = 12$ home runs in a 600 at-bat season. If this value decreases to 40 at bats per home run, we expect $600/40 = 15$ home runs; if it decreases to 30 we expect 20 home runs, and if it decreases to 20 we expect 30 home runs. Thus, the practical importance of a change of 10 units in the at bats per home run measure changes throughout the range of the variable. This analysis suggests that a home run per at bat is the more appropriate way to measure home run performance.

A related issue is the distribution of the performance measure for the players under consideration. Figure 2.9 contains a histogram of home runs per at bat for 2011 MLB players with a qualifying number of plate appearances (Dataset 2.6). Figure 2.10 contains a similar histogram for at bats per home run. Clearly, the distribution of home runs per at bat is more symmetric, and closer to being approximately bell shaped, than is the distribution of at bats per home run, supporting our conclusion that home runs per at bat is the better measure of home run performance. Also, note that, because a player could hit zero home runs in a season, there is no upper limit to the possible values of at bats per home run; hence, the histogram contains a "More" category for very large values.

Therefore, based on the analysis here, home runs per at bat is a more useful measure of home run production than is at bats per home run. One drawback of home runs

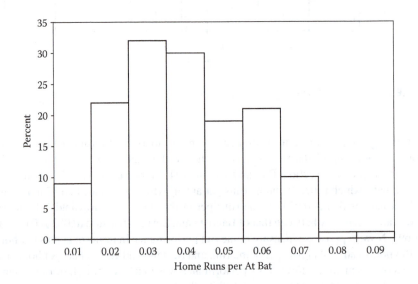

**FIGURE 2.9**    Home runs per at bat for 2011 qualifying MLB players.

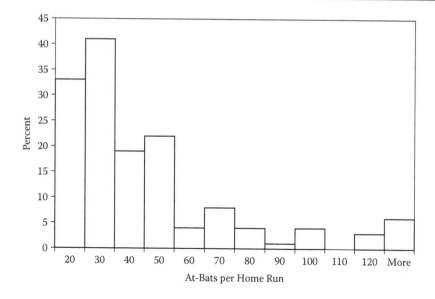

**FIGURE 2.10**    At bats per home run for 2011 qualifying MLB players.

per at bat is that the values tend to be small, making them inconvenient to use. This is easily corrected by using home runs per 100 at bats (or something similar). This is simply home runs per at bat times 100. Therefore, a histogram of these data for 2011 MLB players would be identical to the histogram in Figure 2.9 except that the *x*-axis would have the values 1, 2, and so on instead of 0.01, 0.02, and so on.

# 2.10 COMPUTATION

Frequency tables for discrete data are easily constructed in Excel using the "COUNTIF" function. For example, column A in the Excel worksheet in Figure 2.11 contains the game-by-game results for the Seattle Seahawks' 2012 season, with "W" denoting win and "L" denoting loss. To count the number of wins, we can use the function

*COUNTIF(A1..A16, "W");*

this returns the value 11. Repeated use of this function yields the frequency table.

For continuous data, for which the frequency table is defined in terms of ranges, the histogram procedure found in the Data Analysis package of Excel can be used to construct the frequency table. Consider the data on Tom Brady's by-game passing yards in the 2001–2011 NFL seasons, analyzed in Section 2.3. Figure 2.12 contains a screenshot of the first several rows of the spreadsheet containing those data; the full data are in rows 1 through 159.

| | A | B |
|---|---|---|
| 1 | L | |
| 2 | W | |
| 3 | W | |
| 4 | L | |
| 5 | W | |
| 6 | W | |
| 7 | L | |
| 8 | L | |
| 9 | W | |
| 10 | W | |
| 11 | L | |
| 12 | W | |
| 13 | W | |
| 14 | W | |
| 15 | W | |
| 16 | W | |
| 17 | | |
| 18 | | |

**FIGURE 2.11**    Spreadsheet of Seahawks' game-by-game wins and losses in 2012.

The dialog box for the histogram procedure is given in Figure 2.13. The input range is simply the range of the data to be analyzed, in this case A1..A159; this can be entered manually or by highlighting the relevant cells. For the bin range, we enter the location of cells containing an increasing sequence of values that define the ranges for the table. For instance, to construct the frequency table in Table 2.4, the bin values 100, 200, 300, 400, 500, and 600 are used; Figure 2.14 shows these values entered in the spreadsheet, and Figure 2.15 shows the histogram dialog box with the input values included.

The results of the histogram procedure are given in Figure 2.16. Here, "bin" refers to the range of values greater than the previous bin value up to and including the present bin value. For example, for the results in Figure 2.16, the bin value 200 refers to the range 101–200. The first bin value refers to the range corresponding to the minimum value (or some logical minimum as in the passing yards example) to the present bin value; the final category is always "more," and it includes all data values greater than the largest bin value. Table 2.4 contains the results obtained from Figure 2.16 in a more familiar table form. Note that the histogram procedure can also be used for discrete, quantitative data, but it cannot be used for qualitative data.

| ◢ | A | B | |
|---|---|---|---|
| 1 | 168 | | |
| 2 | 86 | | |
| 3 | 364 | | |
| 4 | 202 | | |
| 5 | 203 | | |
| 6 | 250 | | |
| 7 | 107 | | |
| 8 | 185 | | |
| 9 | 258 | | |
| 10 | 213 | | |
| 11 | 218 | | |
| 12 | 237 | | |
| 13 | 108 | | |
| 14 | 198 | | |
| 15 | 294 | | |
| 16 | 269 | | |
| 17 | 410 | | |

**FIGURE 2.12**   First several rows of the spreadsheet with Brady's data.

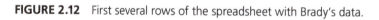

**Histogram**   ? ×

Input

Input Range: ☐

Bin Range: ☐

☐ Labels

Output options

○ Output Range: ☐

◉ New Worksheet Ply: ☐

○ New Workbook

☐ Pareto (sorted histogram)
☐ Cumulative Percentage
☐ Chart Output

OK

Cancel

Help

**FIGURE 2.13**   The dialog box for the histogram procedure.

| | A | B |
|---|---|---|
| 1 | 168 | 100 |
| 2 | 86 | 200 |
| 3 | 364 | 300 |
| 4 | 202 | 400 |
| 5 | 203 | 500 |
| 6 | 250 | 600 |
| 7 | 107 | |
| 8 | 185 | |
| 9 | 258 | |
| 10 | 213 | |
| 11 | 218 | |

**FIGURE 2.14**   The bin endpoints in the spreadsheet.

To construct a histogram, there are two options. For continuous data, we can simply check the "Chart output" box in the histogram dialog box (see Figure 2.15). The histogram obtained by this procedure is shown in Figure 2.17; this can now be customized using the usual graphical options available in Excel. However, even without such customization, the basic features of the distribution are apparent. Note that the variable on the y-axis is "frequency" rather than "relative frequency" as in the examples given in this chapter. Alternatively, a histogram can be constructed as a bar chart based on bin midpoints and the corresponding relative frequency; this is the approach used for the histograms displayed in this chapter.

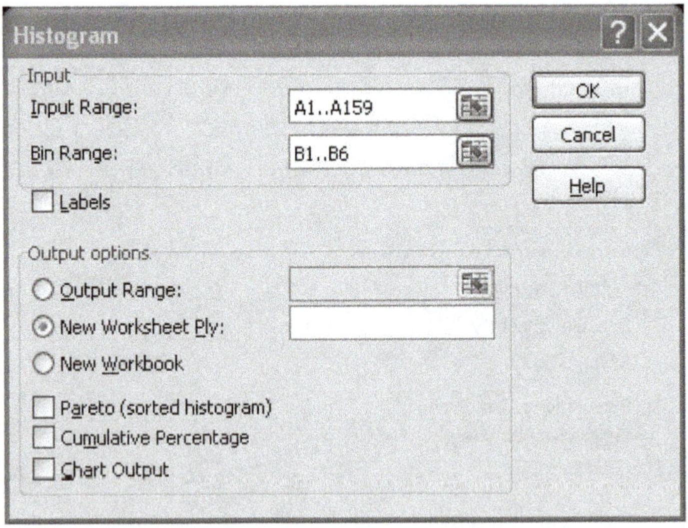

**FIGURE 2.15**   The histogram dialog box with input data for the Brady example.

| | A | B |
|---|---|---|
| 1 | Bin | Frequency |
| 2 | 100 | 5 |
| 3 | 200 | 37 |
| 4 | 300 | 72 |
| 5 | 400 | 42 |
| 6 | 500 | 2 |
| 7 | 600 | 1 |
| 8 | More | 0 |

**FIGURE 2.16**   Results of the histogram procedure.

For descriptive statistics based on quantitative data, such as mean, median, and standard deviation, the "Descriptive Statistics" procedure in the Data Analysis package is useful. The dialog box for this procedure is given in Figure 2.18. The data range is entered in the usual way; it is useful to note that the input range can include several columns, with results given for each column (assuming the "Grouped by Columns" is checked). To obtain the descriptive statistics for these data, be sure to check the "Summary statistics" box; also, if a column label is included in the data range (useful for keeping track of results), you need to check the "Labels in First Row" box.

Figure 2.19 contains the output from the Descriptive Statistics procedure for the batting averages (BA) for the 2011 MLB players with a qualifying number of at bats; the mean and median of those values were reported in Table 2.12. The mean, median, and standard deviation are the statistics described in this chapter. Other useful statistics

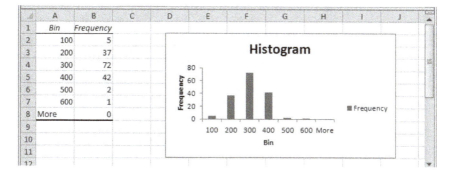

**FIGURE 2.17**   Default histogram of Brady's passing data.

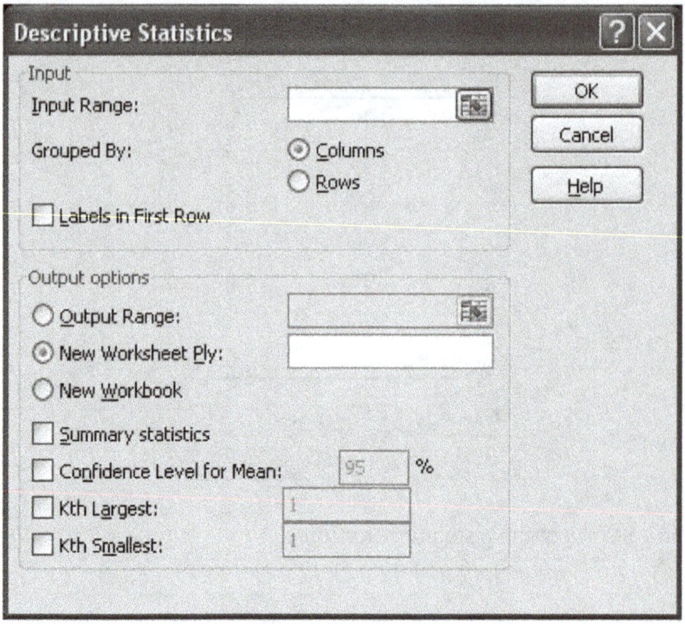

**FIGURE 2.18**   The dialog box for the Descriptive Statistics procedure.

| | A | B |
|---|---|---|
| 1 | BA | |
| 2 | | |
| 3 | Mean | 0.27223 |
| 4 | Standard Error | 0.00232 |
| 5 | Median | 0.26900 |
| 6 | Mode | 0.26200 |
| 7 | Standard Deviation | 0.02798 |
| 8 | Sample Variance | 0.00078 |
| 9 | Kurtosis | -0.41410 |
| 10 | Skewness | 0.26230 |
| 11 | Range | 0.12600 |
| 12 | Minimum | 0.21800 |
| 13 | Maximum | 0.34400 |
| 14 | Sum | 39.47300 |
| 15 | Count | 145.00000 |
| 16 | | |
| 17 | | |

**FIGURE 2.19**   Output from the Descriptive Statistics procedure.

are the maximum and minimum, which are simply the maximum and minimum of the input values; these are useful for detecting errors in the data. The range is the difference of the maximum and minimum, and the value for count indicates the number of observations in the dataset.

The quartiles and the IQR are not included in the output from the Descriptive Statistics procedure. However, these are easily calculated using the "quartile" function in Excel. For instance, to calculate the upper quartile of the data in the range $A1..A159$, use the command

$QUARTILE(A1..A159, 3);$

the lower quartile is calculated using

$QUARTILE(A1..A159, 1).$

Excel has a large number of built-in functions that can be used for transformations. In particular, $LN$ is used to calculate natural logarithms, as discussed in Section 2.8. To calculate $pLN(p)$, as used in calculating the entropy discussed in Section 2.7, it is convenient to use $pLN(p + d)$, where $d$ is a very small number, such as 0.000001. Then, $pLN(p + d)$ is 0 when $p = 0$; for $p > 0$, the effect on the result is negligible.

# 2.11 SUGGESTIONS FOR FURTHER READING

The topics covered in this chapter are standard topics in introductory statistics; therefore, they are covered in many books. My favorite is that of Agresti and Finlay (2009), which discusses a wide range of statistical topics at a level suitable for beginners; also, it emphasizes social science applications, which are closely related to applications in sports. Other good introductory texts include those of McClave and Sincich (2006) and Moore and McCabe (2005). The work of Snedecor and Cochran (1980) is a classic text on statistical methods. Its format is a bit dated, but it has a wealth of useful information.

Huff's (1993) work contains an entertaining introduction to many statistical concepts with an emphasis on graphical methods and taking the point of view of teaching the reader "how to lie with statistics." Tufte (2001) gives a comprehensive discussion of statistical graphics and methods of displaying data.

The analysis of within-team and between-team variation, discussed in Section 2.6, is an example of a general methodology in statistics, known as *analysis of variance* (ANOVA). See, for example, the work of Snedecor and Cochran (1980, Chapter 13) for further discussion and more general results.

# Probability

---

## 3.1 INTRODUCTION

---

The goal of analytic methods is to use data to better understand the factors that influence the results of sporting events. However, to do this, we must deal with the fact that all sports have a random element that comes into play; understanding this randomness is crucial to being able to draw the appropriate conclusions from sports data.

Probability theory is the branch of mathematics that deals with random outcomes. This chapter considers the basic properties and rules of probability, focusing on those concepts that are useful in analyzing sports data.

---

## 3.2 APPLYING THE RULES OF PROBABILITY TO SPORTS

---

The starting point for probability theory is the concept of an *experiment*. An experiment is simply any process that generates a random outcome. For example, Derek Jeter batting against Josh Beckett, Kobe Bryant shooting a free throw, and the Packers playing the Bears are all examples of this type of experiment. The specification of an experiment might be fairly general (e.g., "Jeter bats against Beckett") or detailed (e.g. "Jeter bats against Beckett in the first inning of a scoreless afternoon game in August 2012 at Yankee Stadium with a runner on first and no outs"), depending on the goals of the analysis.

When analyzing an experiment, we are interested in specific "events." An event is anything that might occur as a result of the experiment. The only requirement is that, based on the result of the experiment, we know whether the event has occurred. For instance, in the Jeter-Beckett example, "hit," "home run," "strikeout," and "ground out" are all events; however, "Yankees win" would not qualify as an event because, once the Jeter-Beckett at bat occurs, we do not know if the Yankees win.

Corresponding to each event of an experiment is its probability. If $A$ is a possible event in an experiment, it is often useful to denote the probability of $A$ by $P(A)$. We can think of $P(\cdot)$ as a function that takes different events and returns their probability. For example, in the Jeter-Beckett example, if $H$ denotes "hit," we might write $P(H) = 0.3$ to indicate that the probability that Jeter gets a hit is 0.3.

Although everyone has a good intuitive notion of what a probability is, to use probabilities in a formal analysis we need to have a precise definition. The probability of an event is usually defined as a "long-run relative frequency"; for example, when we say that $P(H) = 0.3$ in the Jeter-Beckett example, we mean that if Jeter faces Beckett in a long sequence of at bats, he will get a hit about 30% of the time.

Note that this idea is a hypothetical one. For instance, consider the experiment that the Bears play the Packers at home; if we say that the probability that the Bears win is 0.25, we mean that in a long sequence of such games the Bears would win about 25% of the time. However, it is impossible to play a long sequence of games under the same conditions; players would age, some would be injured, coaches would change strategy based on what worked and what did not work, and so on. Thus, the "long-run relative frequency" interpretation is just a way to think about probabilities, not a procedure for determining them.

Probabilities follow some basic rules. Consider an experiment and let $A, B$ be events with probabilities, $P(A)$, $P(B)$, respectively. It is often useful to refer to results in which either $A$ or $B$ occurs, which we write "$A$ or $B$". If $A, B$ cannot occur simultaneously, then

$$P(A \text{ or } B) = P(A) + P(B).$$

For instance, in the Jeter-Beckett example, let $S$ denote the event that Jeter hits a single and let $D$ denote the event in which Jeter hits a double. If $P(S) = 0.2$ and $P(D) = 0.05$, then the probability that he hits either a single or a double is

$$P(S \text{ or } D) = P(S) + P(D) = 0.2 + 0.05 = 0.25.$$

Let $H$ denote the event in which Jeter gets a hit (of any type) and suppose $P(H) = 0.3$. Note that because a single is a hit, $S$ and $H$ can occur simultaneously; hence, we cannot calculate $P(S \text{ or } H)$ by summing $P(S)$ and $P(H)$.

A simple, but surprisingly useful, rule applies when we are interested in the probability that an event $A$ *does not* occur; we denote such an event by "not $A$". For example, in the Jeter-Beckett example, "not $S$" is the event that Jeter does not hit a single. If the probability that Jeter hits a single is 0.20, then the probability that he does not hit a single must be 0.80. In fact, this is correct; in general,

$$P(\text{not } A) = 1 - P(A).$$

The likelihood of an event is most commonly expressed in terms of its probability, as we have done so far in this section. However, in some cases, it is more convenient to use *odds* rather than probabilities. Consider an event $A$ with probability $P(A)$. The odds of $A$ occurring or, equivalently, the odds in favor of $A$, are given by the ratio of the probability that $A$ occurs to the probability that $A$ does not occur,

$$\frac{P(A)}{1 - P(A)}.$$

For example, if $A$ has probability 0.5, then the odds in favor of $A$ are 1 to 1. If $A$ has probability 0.75, then the odds of $A$ are 3 to 1; often, we drop the "to 1" in the statement and say simply that the odds of $A$ are 3. Note that we could also talk about the odds against an event, as is often done in gambling; such odds are given by $[1 - P(A)]/P(A)$. It is often convenient to use odds against an event when discussing events that have very small probability.

There are several reasons why odds may be more convenient to use than probabilities. One is that, for very large or very small probabilities, odds are often easier to understand. For instance, if an event has probability 0.00133, in describing the likelihood of this event, it might be more meaningful to say that the odds against the event are about 750 to 1; that is, the event occurs about once in every 750 experiments. Another reason is that, when making a statement about the relative probabilities of events, odds are often easier to interpret. For instance, the statement that the probability of one event is two times the probability of another can mean very different things if the smaller probability is 0.01 or if it is 0.5; furthermore, such a relationship is impossible if the smaller probability exceeds 0.5. On the other hand, if the odds in favor of an event double, the interpretation tends to be more stable over the range of possible values.

Finally, in the context of sports, odds often better represent the relative difficulty of certain achievements. For instance, consider an NFL (National Football League) quarterback and let $A$ denote the event that he throws an interception on a given pass attempt. Based on 2012 data, for Mark Sanchez, $P(A) = 0.040$; for Sanchez to decrease this by 0.01, he would need to throw 9 interceptions in every 300 attempts. For Tom Brady, the probability of $A$ is 0.013. For Brady to decrease this by 0.01, he would need to be almost perfect, throwing only about 1 interception in every 300 attempts. In terms of odds, changing $P(A)$ from 0.04 to 0.03 is equivalent to changing the odds against an interception from 24 to 49. Changing $P(A)$ from 0.013 to 0.003 is equivalent to changing the odds against an interception from about 76 to about 333. Therefore, the odds better reflect the fact that, practically speaking, the difference between interception probabilities of 0.04 and 0.03 is "larger" than the difference between interception probabilities of 0.013 and 0.003.

In this section, we have written the probability of an event by defining a letter, such as $A$, that represents the event and writing $P(A)$ for the probability of $A$. For instance, in the previous example, $A$ represents the event that Mark Sanchez throws an interception, and we stated that, based on 2012 data, $P(A) = 0.040$. Although writing probabilities in this way is useful when describing general rules and formulas for probabilities, in other cases, a more informal approach is preferable. For instance, in the Sanchez example, we might simply write that

$$P(\text{Sanchez throws an interception}) = 0.040,$$

without defining the event $A$. Both approaches are useful, and both are used in the remainder of the book.

# 3.3 MODELING THE RESULTS OF SPORTING EVENTS AS RANDOM VARIABLES

The basic rules of probability are concerned with events, which can describe any possible specific outcome that might occur when an experiment is performed. However, in applying analytic methods to sports, we are generally concerned with data; that is, we analyze numbers, not events. Random variables provide the mathematical link between probability theory and data.

A random variable is simply a numerical quantity derived from the outcome of an experiment. Consider the Bears-Packers example and let $X$ denote the number of touchdown passes thrown by Aaron Rodgers; $X$ is an example of a random variable. Once the outcome of an experiment is available, that is, once the game is played, the value of $X$ for that experiment is known.

A random variable can be used to define events. For instance, in the example, "Rodgers throws one touchdown pass" is an event; in terms of $X$, it can be written "$X = 1$". Therefore, the probability $P(X = 1)$ has the usual interpretation of a long-run frequency. Obviously, there is nothing special about the value 1 in this context, and we might consider $P(X = x)$ for any possible $x$; note that here $X$ denotes a random variable and $x$ denotes a possible value of $X$. The set of values $P(X = x)$ for all possible $x$ is called the *probability distribution* of the random variable.

For instance, in the example, $X$ might follow the distribution given in Table 3.1; the values in this table roughly correspond to Rodgers's career regular season statistics for the games he started through the 2012 season. Note that the probabilities sum to 1, a requirement for a probability distribution.

Events based on random variables, and their probabilities, follow the same rules as other events. Thus, in the example,

$$P(X \leq 1) = P(X = 0 \text{ or } X = 1) = P(X = 0) + P(X - 1) = 0.10 + 0.25 = 0.35.$$

The *distribution function* of a random variable $X$ is the function of $x$ given by $P(X \leq x)$; it is often denoted by $F(x)$. Using the probability distribution in Table 3.1, the corresponding

**TABLE 3.1**  An example of a probability distribution

| $x$ | $P(X = x)$ |
|---|---|
| 0 | 0.10 |
| 1 | 0.25 |
| 2 | 0.25 |
| 3 | 0.25 |
| 4 | 0.10 |
| 5 | 0.05 |

**TABLE 3.2**   An example
of a distribution function

| x | F(x) |
|---|------|
| 0 | 0.10 |
| 1 | 0.35 |
| 2 | 0.60 |
| 3 | 0.85 |
| 4 | 0.95 |
| 5 | 1 |

distribution function is given in Table 3.2. That is, the distribution function is simply the running totals of the probability distribution.

Random variables, in which the set of possible values may be written as a list, for example, 0, 1, 2,... are said to be *discrete*. Hence, the random variable representing Rodgers's touchdown passes is discrete. The probability distribution and distribution function of a discrete random variable can be given as tables, as was done in the example. A *continuous* random variable is one that can take any value in a range. Note that the definitions of discrete and continuous random variables are analogous to the definitions of discrete and continuous data given in Chapter 2.

It is a little more complicated to describe the probability distribution of a continuous random variable. A useful device for expressing such a distribution is to consider a long-run sequence of experiments. If $X$ is a random variable defined for that experiment, there is a corresponding sequence of random variables $X_1, X_2, \ldots$ such that $X_j$ is based on the $j$th experiment. Consider computation of a histogram based on $X_1, \ldots, X_n$, where $n$ is some very large number. Because $n$ is large, such a histogram could be expressed as a smooth curve such as in Figure 3.1. This function gives a description of the probability distribution of $X$. Values of $x$ for which the function is large are relatively more likely than values of $x$ for which the function is small.

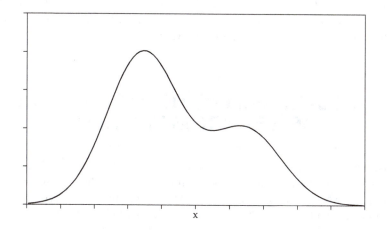

**FIGURE 3.1**   Hypothetical histogram.

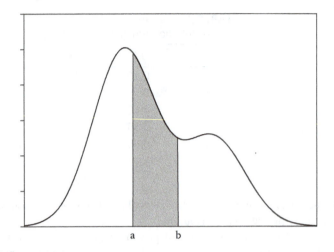

**FIGURE 3.2**   $P(a < X < b)$.

Suppose that the function in Figure 3.1 is standardized so that the total area under the curve is 1. Then, the function can be used to express probabilities regarding $X$. Specifically, for $a < b$, $P(a < X < b)$ can be represented by the area under the curve between points $a$, $b$ as in Figure 3.2. In fact, such an approach can be made precise by relating the function in the histogram to the "probability density function" of the random variable and calculating areas using calculus; however, that level of rigor is not required for us.

One consequence of this approach to continuous random variables is that, for any choice of $a$, $P(X = a)$ is the area under the curve at $a$. Because, in mathematics, the area of a line is always 0, $P(X = a) = 0$ for any $a$. This is sometimes viewed as a paradox because $P(X = a) = 0$ for all $a$, yet $X$ must take *some* value. However, it is a reasonable result if we keep in mind the long-run relative frequency interpretation of probability. Thus, $P(X = 1) = 0$ does not mean that $X = 0$ is absolutely impossible, only that the relative frequency of $X = 1$ is approximately 0 in a long sequence of experiments. If $X$ can take any value in a range, then any particular value will not be observed often.

# 3.4 SUMMARIZING THE DISTRIBUTION OF A RANDOM VARIABLE

The distribution of a random variable can be complex and contain much information. Hence, it is often useful to summarize probability distributions by a few numbers, in the same way that we summarized a set of data in Chapter 2. In fact, any type of summary that can be applied to a set of data can be applied to a random variable by considering random variables $X_1$, $X_2$,... obtained from a long sequence of experiments. For instance, the mean of a random variable $X$ can be viewed as the limiting value of the sample mean based on a long sequence of repetitions of the experiment. Therefore, the relationship between the mean of a random variable and a sample mean is the same as the relationship between a probability and a sample frequency. The mean of a random variable is sometimes called its *expected value* or its *expectation*.

Consider the example in which $X$ denotes the number of touchdown passes thrown by Aaron Rodgers; using the distribution in Table 3.1, the mean or expected value of $X$ is 2.15. Therefore, according to this result, in a large number of Bears-Packers games, we expect Rodgers to throw about 2.15 touchdown passes per game.

Note that such an expected value is a function of a random variable's probability distribution; for example, in the Rodgers example, the mean of $X$ can be determined by assuming that, in a large sequence of games, the number of touchdown passes thrown by Rodgers follows the relative frequencies given by the probabilities in Table 3.1. To obtain the expected value, we multiply the probabilities by the corresponding value of $X$ and sum the results:

$$2.15 = 0.10(0) + 0.25(1) + 0.25(2) + 0.25(3) + 0.10(4) + 0.05(5).$$

However, we are more interested in interpreting and using expected values than calculating them from given probability distributions.

It is sometimes convenient to denote the mean of a random variable $X$ by $E(X)$. For instance, in the example in which $X$ denotes the number of touchdown passes thrown by Rodgers, $E(X) = 2.15$. This notation is particularly convenient when describing properties of means. For instance, if $X$ and $Y$ are random variables, then

$$E(X+Y) = E(X) + E(Y);$$

that is, the average value of a sum of two random variables is simply the sum of the average values. For example, let $Y$ denote the number of rushing touchdowns scored by Rodgers in a given game and assume that $E(Y) = 0.23$; this corresponds to Rodgers's career average through the 2012 season. Then, Rodgers's average total number of touchdowns is $2.15 + 0.23 = 2.38$.

More generally, if $a$, $b$, and $c$ are constants, then

$$E(aX + bY + c) = a\,E(X) + bE(Y) + c.$$

For instance, in the Rodgers example, let $P = 6\,X + 6\,Y$ denote the total number of points resulting from his touchdown passes and rushing touchdowns. Then,

$$E(P) = 6\,E(X) + 6\,E(Y) = 6\,(2.15) + 6\,(0.23) = 14.28;$$

note that here $c$ is 0.

The same approach used to define the mean of a random variable can be used to define the median, standard deviation, and variance of a random variable. These quantities can all be calculated from a random variable's probability and, hence, can be considered to be a summary of that distribution. For instance, in the example, it may be shown that $X$ has median 2, standard deviation 1.31, and variance 1.73.

Although all features of a random variable can be determined from its probability distribution, it is often convenient to model such summaries of a distribution directly, without explicitly modeling the distribution itself. The reason for this is that a probability distribution may be too detailed a description of a random variable, and it may be easier to focus on those aspects of the random variable that are more relevant for a given analysis. For instance, in the example, we might analyze the mean value of $X$ directly without specifying the details of its distribution.

The same ideas that apply to individual random variables also apply to sets of random variables and to quantities calculated from the random variables. For example, let $X$ denote the innings pitched by CC Sabathia in a given start. Using the ideas in this section, $X$ can be modeled as a random variable. Let $X_1, X_2, X_3, X_4$ denote the values of $X$ for four of Sabathia's starts. Then, each of $X_1, X_2, X_3, X_4$ is a random variable, and the set of values $(X_1, X_2, X_3, X_4)$ may be viewed as an ordered set of random variables. Functions of $X_1, X_2, X_3, X_4$ are also random; for example, $(X_1 + X_2 + X_3 + X_4)/4$, the average of the $X$ values, may be viewed as a random variable.

Sets of random variables also arise when we are interested in several aspects of an experiment. For instance, in the Sabathia example, in addition to $X$, the innings pitched in a given start, we might be interested in $K$, the number of strikeouts, and $W$, the number of walks. Then, $(X, K, W)$ describes certain features of Sabathia's performance. In these cases, we are often interested in the relationships among the variables; methods for analyzing such relationships are the focus of Chapters 6 and 7.

Although the probability theory for sets of random variables quickly becomes complicated, the basic ideas are essentially the same as when dealing with a single random variable.

# 3.5  POINT DISTRIBUTIONS AND EXPECTED POINTS

A type of probability distribution that is particularly important in analyzing sports contests is the distribution of the points, or runs, scored in a given scenario. Although the basic idea can be applied to any sport, for concreteness, here we discuss baseball. For example, consider a baseball game in which a team has a runner on first base with 2 outs. Let $X$ denote the number of runs scored by the batting team in the remainder of the inning. We can model $X$ as a random variable with a specific probability distribution. That distribution gives the probability that the team scores 0 runs, 1 run, and so on. Such probabilities are useful for evaluating potential strategies as well as measuring the impact that a certain play has on the outcome of the game.

For the scenario given, Tango, Lichtman, and Dolphin (2007, Table 8) state that $P(X = 0) = 0.865$, $P(X = 1) = 0.063$, $P(X = 2) = 0.050$, $P(X = 3) = 0.016$, $P(X = 4) = 0.005$, and $P(X \geq 5) = 0.003$. These values, taken together, give us the probability distribution of $X$.

For many purposes, the full distribution of runs scored has more detail than needed. A useful summary is the expected value of $X$, as discussed in the previous section. This expected value is known as the expected runs of a given scenario. For the case considered in this section, it can be shown that the expected runs is 0.244; all expected runs values and runs distributions in this chapter are based on the work of Tango, Lichtman, and Dolphin (2007, Table 8).

The expected runs value gives an easily interpreted evaluation of any "state" (number of outs and the position of any base runners) of a baseball game. For instance, if a team has a runner on second base with 2 outs, the expected runs is 0.343. Therefore, moving the runner from first to second, for example, by stealing second, corresponds to an increase of about 0.1 expected runs, or about 1 run for every 10 times the runner moves from first to second with 2 outs.

Although expected runs are useful in analyzing baseball games, there are some important issues to keep in mind. One is that, like any expected value, the expected value of runs scored is a single-number summary of a probability distribution, and in some cases more information regarding the distribution might be important.

For instance, let $X_1$ denote the runs scored in an inning with a runner on third base with no outs and let $X_2$ denote the runs scored in an inning with runners on second and third base with 1 out. The expected value of $X_1$ is 1.445, and the expected value of $X_2$ is 1.451; so, by the expected runs criterion, these two situations are nearly identical. However, $P(X_1 = 0) = 0.143$ while $P(X_2 = 0) = 0.302$, so that, in the second scenario, the team is about twice as likely to be held scoreless than in the first scenario. On the other hand, $P(X_1 \geq 2) = 0.307$ while $P(X_2 \geq 2) = 0.413$, so that the team in the second scenario is also more likely to score 2 or more runs.

Another issue is the method used to determine the probability distribution of runs scored. There are two basic approaches used. The most common approach is to use vast amounts of historical data to estimate the distribution by looking at those cases in which a given scenario occurs. An alternative is to use a probability model based on the probability of certain outcomes (single, double, and so on) and use those results to determine the probability distribution of runs scored. This second approach is more commonly used in football, in which there are many more possible scenarios (position on the field, down and yards needed for a first down), so that some situations are relatively rare in historical data.

Finally, the probability distributions of runs and expected runs are based on averages over several years and for all teams. Therefore, they do not account for the specifics of a game situation based on particular teams and players.

In spite of any possible drawbacks, the use of expected points is an important general approach that has yielded many insights into baseball as well as other sports.

## 3.6 RELATIONSHIP BETWEEN PROBABILITY DISTRIBUTIONS AND SPORTS DATA

It is often useful to model the results of sporting events as random variables. However, once the game has been played, the results are known and hence no longer random. Therefore, it is important to understand the relationship between these data and the corresponding random variables used for modeling purposes.

Let us look at this issue in the context of a simple example. Consider Albert Pujols's at bats during the 2011 MLB (Major League Baseball) season. Let $X$ denote the outcome of a particular at bat, with $X = 1$ indicating a hit and $X = 0$ representing anything that is not a hit. We can model $X$ as a random variable with some probability distribution. However, the properties of $X$ might depend on the exact situation of the at bat (the pitcher, inning, etc.), so let us view $X$ as the result of a randomly chosen at bat (or "random" situation).

Because $X$ takes only two values, 1 and 0, its distribution is characterized by the probability that $X$ is 1, which we can denote by $\pi$, so that $\pi = P(X = 1)$; here, $\pi$ is some number between 0 and 1. Because if $X$ is not 1 it must be 0, $P(X = 0) = 1 - \pi$.

The value of $\pi$ represents Pujols's "true" batting average, that is, his batting average if we were to observe a large number of at bats. This true batting average is unknown, and will never be known, but it is a useful concept when thinking about Pujols's batting ability. For instance, in predicting Pujols's future performance, we would expect the value of $\pi$ to be important.

We do have some information about $\pi$, obtained by observing Pujols's performance in the 2011 season. In 2011, Pujols had 579 at bats and 173 hits. Thus, the observed values of $X$ for the season are a sequence of ones and zeros, with 173 ones and 406 zeros. Based on these results, the empirical counterpart of $\pi$ is $p = 173/579 = 0.299$, which is simply his 2011 batting average. Note that $\pi$ and $p$ represent different measures of Pujols's batting ability— $\pi$ is his "true" batting average for 2011, the batting average he would obtain under a large number of at bats, while $p$ is his observed batting average based on his 579 at bats in 2011.

Although $\pi$ and $p$ are different, we expect that they will be close because 579 at bats provide a lot of information about his true batting average. In fact, as the number of at bats increases, we expect $p$ to become closer and closer to $\pi$; this property is an example of the *law of large numbers.* This law states that, for a large sample size, empirical quantities such as sample means, relative frequencies, and so on will approach their "true" values. That is, roughly speaking, the law of large numbers tells us that data can be used to learn about the process generating the data, and the larger the sample size, the more information we have about that process.

However, it is important to keep in mind that probability models, like all models, include some idealizations and assumptions that must be considered before trusting the results of the model. For instance, in the Pujols batting example, we have assumed that the probability of a hit is the same in each at bat. In practice, however, this probability would likely vary depending on the specific circumstances, such as the pitcher he is facing, the runners on base, the park in which the game is played, and so on. For some purposes, such as modeling Pujols's yearly batting average, the assumption that all at bats have the same probability of a hit might be reasonable. However, for other purposes, such as trying to predict the result when Pujols faces Sabathia in Yankee Stadium, a more detailed analysis might be necessary.

# 3.7 TAILORING PROBABILITY CALCULATIONS TO SPECIFIC SCENARIOS: CONDITIONAL PROBABILITY

Probabilities are interpreted as long-run relative frequencies in a large sequence of experiments. For example, in the Bears-Packers experiment in which the Bears play the Packers at home, if the event that the Bears win has probability 0.25, this means that in a hypothetical long sequence of games, the Bears will win about 25% of the time. An important part of this interpretation is that the 25% applies to all of the experiments in the sequence. However, in some cases, we might only be interested in those experiments satisfying some further conditions.

Continuing the example, let $B$ be the event that the Bears win so that $P(B) =$ 0.25. Now, consider another event, the one in which Jay Cutler throws 4 interceptions; denote this event by $C$. Suppose we are interested in the Bears' probability of winning in those games in which Cutler throws 4 interceptions. We can describe this probability as "the probability that the Bears win *given that* Cutler throws 4 interceptions"; symbolically, we write this $P(B \mid C)$, where the vertical line is read "given that." Probabilities such as $P(B \mid C)$ are called *conditional probabilities* because they include additional conditions.

Note that, in $P(B \mid C)$, we are interested in the probability of $B$; $C$ simply describes the conditions under which the probability is to be calculated. Thus, if, for example, $P(B \mid C) = 0.05$, then in a long sequence of games *in which Cutler throws 4 interceptions,* the Bears win about 5% of the time. We might write this conditional probability more informally as

*P(Bears beat the Packers | Cutler throws 4 interceptions)* = 0.05.

Conditional probabilities are useful because they allow us to incorporate additional assumptions, or additional information, into the probability calculation. Conditional probabilities can be determined from standard, unconditional probabilities. To see how this can be done, consider how we would calculate the conditional probability in the example if we had access to the results from the long sequence of games.

We are interested in those games in which Cutler throws 4 interceptions, so if we are reading a sequence of game summaries, one per page; we place those in which Cutler threw 4 interceptions in a separate pile, a $C$ pile. We now go through that pile and put those that the Bears won in a second pile. The probability $P(B \mid C)$ is the ratio of the number of games in the second pile to the number of games in the $C$ pile. Note that the second pile consists of those games that Cutler throws 4 interceptions *and* the Bears win; denote this event by "$B$ and $C$." Thus, in probability terms,

$P(B \mid C) = P(B \text{ and } C)/P(C)$.

Consider the expected points example in Section 3.5 in which $X_1$ denotes the runs scored in an inning with a runner on third base with no outs and $X_2$ denotes the runs scored in an inning with runners on second and third base with 1 out. Recall that, although the expected values of $X_1$ and $X_2$ are close, there are important differences in their probability distributions.

One way to describe the differences between $X_1$ and $X_2$ is to compare their values of $P(X = 1 \mid X \geq 1)$, the conditional probability that the team scores exactly 1 run given that they score at least 1 run. Note that

$$P(X = 1 \mid X \geq 1) = \frac{P(X = 1 \text{ and } X \geq 1)}{P(X \geq 1)}$$

and because $P(X = 1 \text{ and } X \geq 1) = P(X = 1)$,

$$P(X = 1 \mid X \geq 1) = \frac{P(X = 1)}{P(X \geq 1)}.$$

Using this result, $P(X_1 = 1 \mid X_1 \geq 1) = 0.64$ and $P(X_2 = 1 \mid X_2 \geq 1) = 0.41$ so that, in the first scenario, if the team scores, it is highly likely that they will score only 1 run. On the other hand, in the second scenario, if the team scores, it is more likely to score 2 or more runs than it is to score just a single run.

In general, consider an experiment and let $A$, $B$ be events. Let $A$ *and* $B$ be the event in which both $A$ and $B$ occur. Then,

$P(A \mid B) = P(A \text{ and } B)/P(B).$

Note that this relationship can be written

$P(A \text{ and } B) = P(B) P(A \mid B)$

so that it gives a formula for finding the probability that two events both occur. That is, the probability that both $A$, $B$ occur is the probability that $B$ occurs times the probability that $A$ occurs given that $B$ occurs. Note that, because $A$ *and* $B$ and $B$ *and* $A$ mean the same thing, we also have

$P(A \text{ and } B) = P(A) P(B \mid A),$

giving two options for determining $P(A \text{ and } B)$.

The conditional probability $P(A \mid B)$ can be viewed as an "updated" version of the probability, updated to take into account that $B$ occurs. In many cases, this additional information is important in assessing the probability of interest. For instance, in the Bears-Packers example, knowing that Cutler throws 4 interceptions will change the probability that the Bears win. However, in other cases, knowing that $B$ occurs will not change the probability that $A$ occurs. For instance, if $K$ is the event that the Bears kick off to start the game, then it may be reasonable to assume that

$P(B \mid K) = P(B)$;

that is, the fact that the Bears kick off to start the game does not change their probability of winning the game.

Events $A$, $B$ for which $P(A \mid B) = P(A)$ are said to be *independent*. For independent events, knowledge about one of them does not change our probability of the other. By rearranging the formula for $P(A \mid B)$, it can be shown that $A$, $B$ are independent if and only if

$P(A \text{ and } B) = P(A)\, P(B)$.

That is, for independent events $A$ and $B$, the probability that both occur is simply the product of the individual probabilities.

Consideration of conditional probabilities shows that it is important to be aware of the conditions under which a probability is calculated and to pay close attention to the exact probability considered. In particular, care is needed when interpreting evidence presented in the form of conditional probabilities. This issue is illustrated in the example that follows.

In 2011, Andrew McCutchen had 62 extra-base hits; only 12.9% of these were with 2 or more runners on base; "split" data of this type is available from Baseball-Reference.com. Does this suggest that McCutchen does not hit as well with runners on base? Note that, in MLB in general for 2011, 13.9% of extra-base hits occurred with two or more runners on base.

In analyzing the meaning of results like these, it is often useful to express them using probability notation. The experiment here is a McCutchen at bat, and the events of interest are "had an extra-base hit" and "2 or more runners on base." Note that even though 2 or more runners on base is not a direct consequence of McCutchen's at bat, when the at bat occurs, we know the number of runners on base, so we can view it as an event.

According to 2011 data, 12.9% of the time McCutchen has an extra-base hit, there are 2 or more runners on base. In probability notation,

$P(2 \text{ or more runners are on base} \mid McCutchen \text{ has an extra-base hit}) = 0.129$;

it follows that

$P(less \text{ than 2 runners are on base} \mid McCutchen \text{ has an extra-base hit}) = 0.871$

because 87.1% of the time he has an extra-base hit there are 0 or 1 runners on base. Therefore, 12.9% refers to the tendency of there being 2 or more runners on base when McCutchen has an extra-base hit.

Therefore, when McCutchen has an extra-base hit, it is relatively unlikely that there are at least 2 base runners. Note, however, that these values do not directly assess McCutchen's tendency to have an extra-base hit with 2 or more base runners. To do this, we should look at his extra-base hit probability in the two situations. That is, we should compare

*P(McCutchen has an extra-base hit | 2 or more runners on base)*

and

*P(McCutchen has an extra-base hit | less than 2 runners on base).*

In 2011, McCutchen had only 66 at bats with 2 or more runners on base, and in 8 of those at bats he had an extra-base hit. Therefore,

$$P(McCutchen\ has\ an\ extra\ base\ hit\ |2\ or\ more\ runners\ on\ base) = \frac{8}{66} = 0.121;$$

that is, if he comes to bat with 2 or more runners on base, there is a 12.1% chance he will have an extra-base hit (based on 2011 data). He had 506 at bats with either the bases empty or 1 base runner, and he had 54 extra-base hits. Therefore, based on these data,

$$P(McCutchen\ has\ an\ extra\ base\ hit\ |\ less\ than\ 2\ runners\ on\ base) = \frac{54}{506} = 0.107;$$

that is, if he comes to bat with fewer than 2 base runners, there is about a 10.7% chance that he will have an extra-base hit. It follows that in 2011 McCutchen was actually more likely to have an extra-base hit with 2 or more base runners.

The lesson here is that, in making comparisons of this type, it is important to distinguish between the event of interest (having an extra-base hit) and the event defining the relevant situation (2 or more runners on base) and calculate the probabilities accordingly.

# 3.8  RELATING UNCONDITIONAL AND CONDITIONAL PROBABILITIES: THE LAW OF TOTAL PROBABILITY

There is a simple formula relating unconditional and conditional probabilities. Consider the 2013 St. Louis Cardinals. They won 97 games, for a winning "percentage" of 0.599. However, like most MLB teams, they had a higher winning percentage in home games than in road games. At home, they won 54 of 81 games, for a winning percentage of 0.667, while on the road, they won 43 games, for a winning percentage of 0.531.

These results can be expressed in probability notation. Let $W$ be the event that St. Louis wins and let $H$ denote the event that the game is a home game. Then, $P(W) = 0.599$, $P(W | H) = 0.667$, and $P(W | not\ H) = 0.531$. Because the Cardinals play the same number of home games as away games, their overall winning percentage is simply the average of their home and away winning percentages:

$$0.599 = (0.667 + 0.531)/2,$$

or, in probability notation,

$$P(W) = P(H)P(W \mid H) + P(\text{not } H)(P(W \mid \text{not } H),$$

because $P(H)$ and $P(\text{not } H)$ are both 0.5.

This result, relating unconditional and conditional probabilities, is known as the *law of total probability*. Consider an experiment and let $A, B$ denote events. Then, the law of total probability states that

$$P(A) = P(B)P(A \mid B) + P(\text{not } B)P(A \mid \text{not } B).$$

That is, the unconditional probability of an event $A$ can be expressed in terms of a weighted average of the conditional probabilities of $A$ given $B$ and given "not $B$," where the weights depend on the probability of $B$.

In the winning percentage example, each MLB team plays 81 home games and 81 away games, so that each team's overall winning percentage is the average of its home and away winning percentages. That is, the relationship between the unconditional probability of the event $W$ has a simple relationship to the conditional probabilities of $W$ given $H$ and $W$ given "not $H$." Because all teams play 81 home games and 81 away games, the relationship between home and away winning percentages and the team's overall winning percentage is the same for each team. In other cases, the probabilities of the conditioning event may vary for different players or teams, making comparisons more difficult.

Consider the following example. In 2009, Josh Beckett and Johan Santana both had solid years, with similar statistics. In particular, both pitchers had a batting average against (BAA) of .244, with Santana holding a slight edge with a value of .2438 compared to Beckett's .2441. Furthermore, both were much stronger against right-handed batters: Beckett had a BAA of .226 against right-handed batters and a BAA of .258 against left-handed batters, while Santana had a BAA of .235 against right-handed batters and a BAA of .267 against left-handed batters.

Therefore, Beckett's BAA was 9 points lower than Santana's against right-handed batters *and* against left-handed batters. Yet, their overall BAAs were virtually the same, with Santana's slightly *lower*. This surprising result, called *Simpson's paradox* in statistics, can be explained by the fact that Beckett and Santana faced different proportions of right- and left-handed batters. Of the 811 at bats against Beckett, only 43.2% were from the right side; of the 640 at bats against Santana, 71.9% were by right-handed batters.

The relationship between the pitchers' overall BAAs and their side-specific BAAs follows from the law of total probability. Let $H$ denote the event that, in a given at bat, a pitcher allows a hit. Let $R$ denote the event that the batter is right-handed and let $L$ denote the event that the batter is left-handed; note that $L$ is "not $R$." Then, the law of total probability states that

$$P(H) = P(R)P(H \mid R) + P(L)P(H \mid L).$$

Here, $P(H \mid R)$ is a pitcher's BAA versus right-handed batters, and $P(H \mid L)$ is his BAA versus left-handed batters. Using Beckett's statistics, this relationship becomes

$$0.244 = (0.432)(0.226) + (0.568)(0.258)$$

while for Santana,

$$0.244 = (0.719)(0.235) + (0.281)(0.267).$$

Therefore, while Beckett was better than Santana versus both right-handed and left-handed batters, Santana faced more right-handed batters than did Beckett, lowering his overall BAA. It follows that the conditional probabilities, in this case their BAA values versus right- and left-handed batters, give different information about the relative performance of Beckett and Santana than do the unconditional probabilities, their overall BAA values.

---

# 3.9 THE IMPORTANCE OF SCORING FIRST IN SOCCER

---

Conditional probabilities can be used to incorporate additional information in a probability calculation, as discussed in Section 3.7. In this section, we give an example of this to quantify the importance of scoring first in soccer games played in the English Premier League (EPL) in the 2010–2011 through 2012–2013 seasons. The data analyzed here are available on SoccerSTATS.com.

Based on these data, the probability that the home team wins the game is 0.453. Note that this is an average value for the entire league; we can think of it as applying to a game played between two "randomly chosen" EPL teams. We can write this result as

$P(Home\ team\ wins) = 0.453.$

Now, suppose that we include the additional information that the home team scores first. This additional information changes the probability that the home team wins the game. For the EPL, the conditional probability that the home team wins given that the home team scores first is 0.718. We write this as

$P(Home\ team\ wins \mid Home\ team\ scores\ first) = 0.718.$

Therefore, the fact that the home team scores first has an important effect on the probability that the home team wins the game, changing it from the unconditional probability of 0.453 to 0.718. If the visiting team scores first, the probability that the home team wins is 0.178,

$P(Home\ team\ wins \mid Visiting\ team\ scores\ first) = 0.178.$

Again, the fact that the visiting team scores first is important information that greatly affects the probability that the home team wins. There is the possibility that neither team scores first—that is, there are no goals scored—in this case, the game is a draw so that

*P(Home team wins | Neither team scores first) = 0.*

These three conditional probabilities are related. The probability that the home team scores first is 0.534, the probability that the visiting team scores first is 0.390, and the probability that neither team scores first is 0.076. The law of total probability discussed in the previous section can be extended to apply to three conditioning events:

*P(Home team wins) = P(Home scores first) P(Home team wins | Home scores first)*

  *+ P(Visitor scores first) P(Home team wins | Visitor scores first)*

  *+ P(Neither scores first) P(Home teams wins | Neither scores first).*

Using the probability values based on the EPL, this expression becomes

$$0.453 = (0.718)(0.534)+(0.178)(0.390)+(0)(0.076).$$

Note that most of the home team's wins come in games in which they have scored first. To express this idea, we may write

*P(Home team scores first | Home team wins) = 0.847.*

As discussed in the previous section, this probability is fundamentally different from the probability

*P(Home team wins | Home team scores first) = 0.718.*

The probability 0.847 refers to the fact that in 84.7% of the games in which the home teams wins, the home team scores first. That is, it refers to a proportion of games in which the home team wins. The probability 0.718 refers to the fact that in 71.8% of the games in which the home team scores first, the home team wins the game. That is, it refers to a proportion of games in which the home team scores first.

## 3.10 WIN PROBABILITIES

The analysis in Section 3.9 shows how the probability of an event, such as the home team winning a soccer game, can be updated to reflect additional information, such as the fact that the home team scored the first goal of the game, using conditional probabilities. An extension of this idea is the concept of "win probabilities," which give the probability of a team winning a game as a function of the current status of the game. Although the basic approach can be applied to any sport, here we consider it in the context of football.

Consider an NFL game; because the win probabilities are based on league-wide results and do not take into account the properties of specific teams, let us call the teams Team A and Team B. Before the game starts, the probability that Team A wins is 0.50.

Now, suppose that Team A receives the opening kickoff, which results in a touchback so that Team A starts with the ball at their own 20-yard line. This additional information changes the probability that Team A wins. Using the win probability calculator on the Advanced Football Analysis website (http://wp.advancedfootballanalytics.com/winprobcalc1.php), that probability is now 0.52. This may be viewed as the conditional probability that Team A wins the game given that the team received the opening kickoff and it is a touchback.

As the game unfolds, the probability that Team A wins the game evolves, depending on the specific outcomes. For instance, if on the first play from scrimmage, Team A has a 30-yard pass completion, taking 10 seconds, then their win probability increases from 0.52 to 0.57. If on the next play, the quarterback is sacked, fumbles, and loses the ball at his own 40, the probability that Team A wins drops to 0.41. These win probabilities can be viewed as the conditional probabilities that a team wins the game given the current status (possibly hypothetical) of the game.

Win probabilities are similar to expected points in that they give us a way to evaluate the effect of individual plays or potential strategic decisions. For instance, we might calculate the "win probability added" of a player by computing the change in win probability resulting from the plays in which he is involved.

The main difference between win probabilities and expected points is that win probabilities are heavily context dependent. For instance, a 50-yard pass completion to the opponent's 20-yard line on first down with 2 minutes remaining in the fourth quarter of a tied game increases the offensive team's win probability from 0.69 to 0.86. That same play, but with 2 minutes remaining in the first quarter, increases the offensive team's win probability from 0.54 to 0.64. Therefore, the fourth quarter play has a much greater effect on win probability, as would be expected.

# 3.11  USING THE LAW OF TOTAL PROBABILITY TO ADJUST SPORTS STATISTICS

The law of total probability provides a method of adjusting statistics for different players or teams to make the statistics directly comparable. Recall the example from Section 3.8 concerning the 2009 performance of Josh Beckett and Johan Santana. Although Beckett's BAA is 0.009 lower than Santana's against both right-handed batters and left-handed batters, their overall BAA values are approximately the same. This was explained by the fact that both pitchers had a lower BAA against right-handed batters and Santana faced many more right-handed batters than did Beckett.

Therefore, we might ask what Beckett's overall BAA would have been had he faced the same proportion of right-handed batters as did Santana. Recall that the equation relating Beckett's overall BAA to his BAA versus right- and left-handed batters is

$$0.244 = (0.432)(0.226) + (0.568)(0.258).$$

Here, 0.266 and 0.258 are Beckett's BAA versus right- and left-handed batters, respectively, and the 0.432 and 0.568 are the proportion of right- and left-handed batters, respectively, that he faced.

To estimate what Beckett's BAA would have been had he faced the same proportion of right-handed batters that Santana faced, we replace 0.432 and 0.568 in this equation, reflecting the fact that 43.2% of the batters he faced were right-handed, by 0.719 and 0.281, respectively, the values for Santana. This yields the result

$$(0.719)(0.226) + (0.281)(0.258) = 0.235.$$

That is, if Beckett had faced the same proportion of right-handed batters as did Santana, his overall BAA would be .235, 9 points lower than Santana's, as would be expected from the fact that Beckett's side-specific BAAs were both 9 points lower than Santana's.

Alternatively, we might compute Beckett's and Santana's predicted overall BAA under the assumption that the proportion of right-handed batters is some "standard" value. For instance, in 2009 approximately 56% of all MLB at bats were from the right side. Therefore, an adjusted BAA can be formed by weighting the side-specific BAAs using the weights 0.56, 0.44. Using these weights, Beckett's adjusted BAA is

$$(0.54)(0.226) + (0.46)(0.258) = 0.241,$$

while Santana's adjusted BAA is

$$(0.54)(0.235) + (0.46)(0.267) = 0.250.$$

These calculations are examples of *subclassification adjustment*, which is sometimes called *direct adjustment*. When the statistic of interest can be interpreted as a probability, it is simply the law of total probability, in which the probabilities of the conditioning event are chosen to be some "standard" values.

This method is not restricted to probabilities; it can be applied whenever the measurement of interest can be formed by weighting the measurements from various subclasses. When these subclass weights vary from subject to subject, we may want to make an adjustment that reduces or eliminates the effect of the variation in the subclass weights.

More formally, let Y denote the measurement of interest and let $Y_1, \ldots Y_m$ denote the measurements for the various subclasses, such that

$$Y = q_1 Y_1 + \cdots + q_m Y_m$$

where $q_1,...,q_m$ are the subclass weights, which sum to 1. For instance, in the BAA example, consider Beckett's statistics. The variable $Y$ represents Beckett's overall BAA, $Y_1$ represents his BAA against right-handed batters, and $Y_2$ represents his BAA against left-handed batters. The weights $q_1$, $q_2$ are the proportion of at bats against right-handed and left-handed batters, respectively; note that here $m = 2$.

To form an adjusted value of $Y$ that accounts for the weights $q_1,...,q_m$, we replace these weights by some "standard" weights $p_1,..., p_m$ in the equation, yielding the adjusted value

$$Y^* = p_1Y_1 + \cdots + p_mY_m.$$

Then, $Y^*$ is the predicted value of $Y$ under the assumption that the subclass weights are $p_1,..., p_m$ rather than $q_1,..., q_m$.

Direct adjustment of this type is available whenever a measurement $Y$ is related to subclass-specific measurements $Y_1,...,Y_m$ through an equation of the form

$$Y = q_1Y_1 + \cdots + q_mY_m$$

and we wish to adjust for the subclass proportions. Thus, it is appropriate for rates, counts, and similar statistics. It would not be appropriate for more general types of statistics such as ratios of summary statistics such as the NFL passer rating.

It is important to keep in mind that the relative values of the adjusted statistics will depend on the values chosen for $p_1,..., p_m$. This is not apparent in the Beckett-Santana comparison because Beckett's BAA is 9 points lower than Santana's against both right- and left-handed batters.

Four other pitchers had a BAA of 0.244 in 2009. Together with Beckett and Santana, the 6 pitchers' BAAs against right- and left-handed batters are given in Table 3.3, along with two adjusted BAAs. "Adjusted 1" uses weights based on the overall proportion of batters who are right- and left-handed, 0.54 and 0.46, as in the analysis for Beckett and Santana previously in this section.

The second adjustment ("Adjusted 2") recognizes the fact that because of platooning and the presence of switch hitters, a right-handed pitcher will not face the same proportion of right-handed hitters that a left-handed pitcher does. Hence, for left-handed pitchers, it gives weight 0.35 to BAA versus left-handed batters and weight

**TABLE 3.3**   Adjusted BAA for pitchers with a BAA of 0.244 in 2009

| PITCHER | THROWS | BAA VS. R | BAA VS. L | ADJUSTED 1 | ADJUSTED 2 |
|---------|--------|-----------|-----------|------------|------------|
| Beckett | R | .226 | .258 | .241 | .242 |
| Santana | L | .235 | .267 | .250 | .246 |
| Floyd | R | .256 | .232 | .245 | .244 |
| Billingsley | R | .229 | .257 | .242 | .243 |
| Wainwright | R | .217 | .275 | .244 | .246 |
| Happ | L | .253 | .216 | .236 | .240 |

0.65 to BAA versus right-handed batters; for right-handed pitchers, it gives equal weight to BAA versus left- and right-handed batters. These values correspond to the proportion of batters from each side for each type of pitcher. Although the two adjusted values are similar, they are different; for example, Santana's adjusted BAA is 14 points higher than Happ's using the first adjustment, but it is only 6 points higher using the second adjustment.

Although the standard weights are generally chosen with some objectivity in mind, they do represent a hypothetical scenario that may or may not be realistic. Therefore, although adjusted values provide useful information, they are not perfect. When there are two subclasses, it may be more informative simply to present the subclass-specific values of the statistic under consideration. For example, in Table 3.3, we might report the right- and left-handed BAAs without any attempt to standardize them into a single value. However, this approach becomes less useful as the number of subclasses increases.

# 3.12 COMPARING NFL FIELD GOAL KICKERS

Consider NFL field goal kickers. In 2011, the top field goal kickers in terms of field goal percentage, the percentage of successful attempts, were Matt Bryant (93.1%) and Connor Barth (92.9%). However, different kickers attempt field goals from different distances, depending on opportunity and coaching decisions. Thus, we might consider adjusting field goal percentage for the distance of the attempts, using the direct adjustment approach from the previous section.

Field goal success rates are commonly recorded for the distances 20 to 29 yards, 30 to 39 yards, 40 to 49 yards, and 50+ yards, which includes all attempts of greater than or equal to 50 yards; here, we ignore attempts of 19 yards or less (Dataset 3.1). For a given kicker, let $F$ denote his overall field goal percentage and let $F_{20}$, $F_{30}$, $F_{40}$ and $F_{50}$ denote his field goal percentages for the four ranges described. Then,

$$F = q_{20}F_{20} + q_{30}F_{30} + q_{40}F_{40} + q_{50}F_{50}$$

where, for example, $q_{20}$ denotes the proportion of his attempts in the range of 20 to 29 yards.

To adjust $F$ for the distances of the attempts, we calculate

$$F^* = p_{20}F_{20} + p_{30}F_{30} + p_{40}F_{40} + p_{50}F_{50}$$

where $p_{20}, p_{30}, p_{40}, p_{50}$ are some specified "standard" values. For instance, to adjust a kicker's data from 2011, we might use the proportion of all NFL field goal attempts in 2011 in the various ranges, so that

$$p_{20} = 0.291, \quad p_{30} = 0.265, \quad p_{40} = 0.306, \quad p_{50} = 0.138.$$

Using these values, it is straightforward to calculate $F^*$ for each kicker. For instance, for Bryant, $F_{20} = 100$, $F_{30} = 100$, $F_{40} = 71.4$, and $F_{50} = 100$ so that his adjusted field goal percentage is

$$(0.291)(100)+(0.265)(100)+(0.306)(71.4)+(0.138)(100) = 91.2.$$

In terms of adjusted field goal percentage, the top 2 kickers, of those with at least 20 attempts and at least 1 attempt in each category, are Scobee (93.9%) and Barth (93.2%); Sebastian Janikowski, who is third, showed the greatest increase, from 88.6% to 93.1%, reflecting the fact that Janikowski tried the most field goals of 50 yards or greater of any NFL kicker in 2011. The worst kicker in adjusted field goal percentage is Lawrence Tynes (71.9%).

# 3.13 TWO IMPORTANT DISTRIBUTIONS FOR MODELING SPORTS DATA: THE BINOMIAL AND NORMAL DISTRIBUTIONS

Although the probability distribution of a random variable can be quite general, subject to some simple requirements such as the set of all probabilities summing to 1, in practice, there are a few distributions that are particularly useful. In this section, we consider two such distributions (binomial and normal) in detail.

Consider an experiment and let $A$ be an event of interest. Define a random variable $X$ such that $X = 1$ if $A$ occurs and $X = 0$ otherwise. Then, $P(X = 1) = P(A)$, which we can denote by $\pi$, where $0 < \pi < 1$. Note that we used this approach in Section 3.6 when analyzing Pujols's batting in 2011.

Let $X_1, X_2 ..., X_n$ be independent random variables, each with the distribution of $X$. Then, $X_1, X_2 ..., X_n$ is a sequence of ones and zeros. Let $S = X_1 + X_2 + \cdots + X_n$. Note that $S$ is simply the number of times that $A$ occurs in the $n$ experiments. It follows that $S$ is a random variable, and its distribution can be determined using the information provided. $S$ is said to have a *binomial distribution* with parameters $n$, $\pi$.

For instance, suppose $n = 2$. To find the probability that $S = 2$, we need to find all combinations of $X_1$ and $X_2$ that yield a sum of 2 and add their probabilities. Because there is only one way to have $S = 2$ (both $X_1$ and $X_2$ must be 1), this case is particularly simple:

$$P(S = 2) = P(X_1 = 1, X_2 = 1).$$

Because $X_1, X_2$ are independent:

$$P(X_1 = 1, X_2 = 1) = P(X_1 = 1) P(X_2 = 1) = \pi^2.$$

Finding the probability that $S = 1$ is a little more complicated because there are two ways to have $S = 1$: $X_1 = 1$, $X_2 = 0$ or $X_1 = 0$, $X_2 = 1$ so that

$$P(S = 1) = P(X_1 = 1, X_2 = 0) + P(X_1 = 0, X_2 = 1)$$
$$= P(X_1 = 1)P(X_2 = 0) + P(X_1 = 0) P(X_2 = 1).$$
$$= 2\pi(1-\pi).$$

Because we must have $P(S=0)+P(S=1)+P(S=2)=1$, $P(S=0)$ must be $(1-\pi)^2$.

For general $n$, the same basic argument works, but the details rapidly become complicated. However, virtually all spreadsheet programs, statistical packages, and many calculators can compute binomial probabilities for given values of $n$ and $\pi$; see Section 3.17 for information on how to compute these probabilities in Excel.

The important assumptions of the binomial distribution are that the results of the individual experiments are independent and that the probability that $X = 1$, i.e. that the event of interest occurs, is the same in each experiment.

The situations in which the binomial distribution applies are quite simple—we identify an event of interest and simply count how often that event occurs. However, it is because of that simplicity that the binomial distribution is so useful. Even when the experiment itself is complicated, we are often interested in relatively simple features of the results. For example, if our experiment is an NFL season, the detailed results of that experiment would fill this book. However, suppose we are interested in whether or not the team with the leading rusher during the regular season wins the Super Bowl. Then the number of seasons in the past 20 years in which the team with the leading rusher wins the Super Bowl can be modeled as a binomial random variable.

The only quantities governing the binomial distribution are $n$, the number of experiments and $\pi$, the probability of the event of interest occurring in a given experiment. Therefore, all the properties of a binomial random variable $S$ are functions of $n$, $\pi$. For instance the mean of $S$ is $n\,\pi$. For example, if we observe 100 experiments and in each one the probability of $A$ is 0.25, we expect 25 occurrences of $A$.

The standard deviation of $S$ is $\sqrt{n\pi(1-\pi)}$. The form of this expression may seem a little strange but, after a little reflection, it should make some sense. Standard deviation is a measure of variation. Suppose $\pi$ is very close to 0. Then $A$ almost never occurs. Therefore, $S$ is almost always 0; that is, there is very little variation in $S$. The same argument applies if $\pi$ is very close to 0, except that $A$ almost always occurs and $S$ is almost always $n$. That is, when $\pi$ is close to either 0 or 1, then the standard deviation should be small. We expect a lot of variation whenever $\pi = 1/2$ because $A$ and "not $A$" are equally likely. The function $\sqrt{n\pi(1-\pi)}$ has all these properties: It is 0 when either $\pi = 0$ or $\pi = 1$, and it takes its maximum value when $\pi = \frac{1}{2}$.

The second important distribution that we will consider is the *normal distribution*. Unlike the binomial distribution, the normal distribution is a continuous distribution, and if a random variable $X$ has a normal distribution, $X$ can take any value between $-\infty$ and $\infty$, although extreme values are unlikely. We briefly discussed the normal distribution in Chapter 2, where it was presented as the "ideal" shape of a histogram.

The normal distribution is governed by two parameters, traditionally denoted by $\mu$ and $\sigma$. Here, $\mu$ represents the mean of the distribution of $X$, and $\sigma$ represents the standard deviation; because standard deviations are always positive, $\sigma > 0$. The shape of the distribution is given by the well-known bell-shaped curve, which takes its maximum value at $\mu$; $\sigma$ governs how spread out the curve is. Figure 3.3 shows several normal distributions, corresponding to different values of $\mu$ and $\sigma$. These plots illustrate some important properties of the normal distribution. For instance, the distribution is symmetric about its peak, which occurs at the mean of the distribution. When the value of $\mu$ changes, the effect on the distribution is a shift; other aspects of the distribution, such as its "bell shape," do not change. When the value of $\sigma$ changes, the effect is essentially to change the scale on the $x$-axis.

Although it is easy to describe the shape of a normal distribution, it is a little more difficult to determine probabilities associated with a normal distribution. Let $X$ denote a random variable with a normal distribution with mean $\mu$ and standard deviation $\sigma$. Since $X$ is a continuous random variable, we cannot give a table listing the possible values of $X$ together with their probabilities. Instead, we consider the probability that $X$ falls into certain ranges of values.

To describe such probabilities, it is useful to relate $X$ to a *standard normal distribution*. A standard normal distribution is the one in which $\mu = 0$ and $\sigma = 1$. Let $X$ denote a random variable with mean $\mu$ and standard deviation $\sigma$. We can convert $X$ to a random variable with a standard normal distribution by computing

$$Z = \frac{(X - \mu)}{\sigma}.$$

Then, $Z$ has a standard normal distribution. The standard normal distribution is a convenient reference distribution that can be used to understand variation in many different contexts; several examples of this are presented in further chapters of this book.

Let $Z$ have a standard normal distribution and consider $P(-a < Z < a)$ as a function of $a$. This probability tells us how close $Z$ tends to be to 0, the mean of its distribution. Table 3.4 gives values of this probability for several values of $a$.

Note that the value of $P(-a < Z < a)$ rapidly approaches 1 as $a$ increases. This is an important property of the normal distribution—with high probability, normal random variables tend to be close to their mean value.

Now, consider a normal random variable $X$ that has mean $\mu$ and standard deviation $\sigma$. To find $P(-b < X < b)$ for some value $b$, we convert this probability into a probability concerning a standard normal random variable $Z$. Because $(X - \mu)/\sigma$ has a standard normal distribution,

$$P(-b < X < b) = P(-(b - \mu)/\sigma < (X - \mu)/\sigma < (b - \mu)/\sigma) = P(-((b - \mu))/\sigma < Z < ((b - \mu))/\sigma);$$

$P(-b < X < b)$ can now be found from Table 3.4 (or a similar table) by using $a = (b - \mu)/\sigma$, which is called the *Z-score of $b$*. The Z-score of $b$ is the number of standard deviations it is above or below the mean of the distribution. The fact that $P(-1 < Z < 1) = 0.683$ can be interpreted as saying that the probability that a normal random variable falls within one

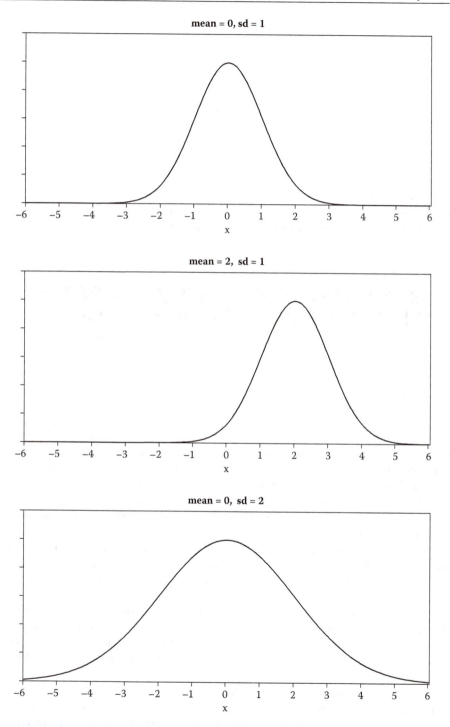

**FIGURE 3.3**   Examples of normal distributions.

**TABLE 3.4** Probabilities for a standard normal random variable

| a | P(–a < Z < a) |
| --- | --- |
| 0.5 | 0.383 |
| 1 | 0.683 |
| 1.5 | 0.866 |
| 2 | 0.954 |
| 3 | 0.997 |

standard deviation of its mean is 0.683. Therefore, in the statement "with high probability, normal random variables tend to be close to their mean value," *close* is interpreted as being relative to the standard deviation of the random variable.

# 3.14 USING Z-SCORES TO COMPARE TOP NFL SEASON RECEIVING PERFORMANCES

The Z-score of a measurement Y is defined as

$$\frac{Y - \mu}{\sigma}$$

where $\mu$ is the mean of Y values for some distribution, and $\sigma$ is the corresponding standard deviation. In the previous section, Z-scores were used as a way to understand, and calculate, probabilities regarding random variables with normal distributions. In this section, Z-scores are used to compare and standardize measurements.

The Z-score gives the number of standard deviations a measurement is above or below the mean. For example, if a measurement has a Z-score of 2, then that measurement is 2 standard deviations greater than the average value. Therefore, the Z-score takes into account both the average value of the measurement and the variability of the measurement, as measured by the standard deviation; Z-scores give a simple way to compare a statistic for a particular player or team to the values obtained by other players or teams.

In 2012, Calvin Johnson had 1964 receiving yards, the highest yearly total up to that time. It is natural to ask how Johnson's 2012 season compares to other great years for receivers. Table 3.5 gives the receiving yard totals for Johnson and 5 other receivers; these were chosen to represent a wide range of eras and are not necessarily the 5 best seasons in terms of receiving yards.

Direct comparison of the six season totals in Table 3.5 may be misleading because of the way the role of the passing game in the NFL has changed over the years. One way to account for these differences is to compare each receiver to the other receivers that played that season.

**TABLE 3.5** Top receiving yard performances in different eras

| *PLAYER* | *YEAR* | *RECEIVING YARDS* |
|---|---|---|
| Calvin Johnson | 2012 | 1964 |
| Marvin Harrison | 2002 | 1722 |
| Jerry Rice | 1995 | 1848 |
| John Jefferson | 1980 | 1340 |
| Otis Taylor | 1971 | 1110 |
| Raymond Berry | 1960 | 1289 |

Using the Z-score approach, we convert each player's performance to a Z-score and then compare the Z-scores of the different players; the player with the highest Z-score has the best performance relative to his peers. Let $Y_0$ be the receiving yards for a player in a given year. For the mean and standard deviation of $Y_0$, we must use the sample-based values. Therefore, let $\bar{Y}_0$ denote the average receiving yards for some group of players and let $S_0$ denote the standard deviation of receiving yards for those players. Then, the Z-score of $Y_0$ is $(Y_0 - \bar{Y}_0) / S_0$.

To implement this approach, we need to choose the players to use to calculate $\bar{Y}_0$ and $S_0$. One possibility is to use the set of all players catching at least one pass in the given year. Table 3.6 gives the average and standard deviation of receiving yards for those players for each year represented in Table 3.5. Table 3.6 also contains the Z-scores of the receiving yards for those players in Table 3.5. For example, for 2002, $\bar{Y}_0 = 291.2$ and $S_0 = 325.8$, so that the Z-score for Marvin Harrison's 1722 receiving yards in 2002 is

$$\frac{1722 - 291.2}{325.8} = 4.392$$

so that Harrison's 1722 yards are about 4.4 standard deviations above the average number of yards for all players with at least 1 reception in 2002.

**TABLE 3.6** Mean and standard deviation of receiving yards for all players with at least one reception for the years represented in Table 3.5

| *YEAR* | *AVERAGE* | *SD* | *Z-SCORE* |
|---|---|---|---|
| 2012 | 269.2 | 329.6 | 5.142 |
| 2002 | 291.2 | 325.8 | 4.392 |
| 1995 | 288.4 | 342.8 | 3.600 |
| 1980 | 280.5 | 278.4 | 3.806 |
| 1971 | 224.3 | 223.8 | 3.958 |
| 1960 | 234.0 | 263.9 | 4.032 |

According to the results in Table 3.6, Calvin Johnson's performance in 2012, which corresponds to a Z-score of about 5.1, is the most impressive, with his receiving yards over 5 standard deviations higher than the average receiving yards for players with at least 1 reception.

The same basic approach can be used to form adjusted statistics that can be compared directly. Suppose we want to determine the value of $Y_0$ for an alternative set of conditions, such as a different era or league. Let $\overline{Y}_1$ and $S_1$ denote the average and standard deviation of this statistic under these alternative conditions; we assume that the values of $\overline{Y}_1$ and $S_1$ are available.

An adjustment using Z-scores is based on the assumption that the Z-scores remain constant under different conditions so that

$$\frac{Y_0 - \overline{Y}_0}{S_0} = \frac{Y_1 - \overline{Y}_1}{S_1}.$$

Therefore, the value of $Y_0$ adjusted to the conditions of $\overline{Y}_1$ and $S_1$ is given by

$$\frac{S_1}{S_0}\left(Y_0 - \overline{Y}_0\right) + \overline{Y}_1.$$

Note that this adjustment has both multiplicative and additive components.

To apply this approach to the receivers' example, we convert each player's performance to a Z-score and then use that Z-score to estimate the player's performance in some "standard" season, which we take to be 2012.

To perform the adjustment, we take each player's Z-score and convert it to the equivalent yardage value for 2012. For example, Harrison's Z-score is 4.392; using the formula for the Z-score in 2012,

$$\frac{Y - 269.2}{329.6}$$

we solve

$$\frac{Y - 269.2}{329.6} = 4.392$$

for $Y$, which gives us the adjusted value of 1717 in "2012 yards." Note that this is equivalent to using the formula given previously in this section:

$$\frac{S_1}{S_0}\left(Y_0 - \overline{Y}_0\right) + \overline{Y}_1$$

with year 1 taken to be 2012 and year 0 taken to be 2002,

$$\frac{S_1}{S_0}\left(Y_0 - \overline{Y}_0\right) + \overline{Y}_1 = \frac{329.6}{325.8}\left(1722 - 291.2\right) + 269.2 = 1717.$$

**TABLE 3.7**   Adjusted receiving yards for players in Table 3.5

| PLAYER | YEAR | YARDS | ADJ-ALL | ADJ-TOP |
|--------|------|-------|---------|---------|
| Johnson | 2012 | 1964 | 1964 | 1964 |
| Harrison | 2002 | 1722 | 1717 | 1881 |
| Rice | 1995 | 1848 | 1456 | 1795 |
| Jefferson | 1980 | 1340 | 1524 | 1811 |
| Taylor | 1971 | 1110 | 1574 | 1832 |
| Berry | 1960 | 1289 | 1598 | 1859 |

Table 3.7 contains the adjusted receiving yards, using this approach, for the receivers and years listed in Table 3.5, under the heading "Adj-All." Note that the value of Johnson's receiving yards, which is already expressed in terms of 2012 yards, does not change. Johnson's 2012 season is the best of those considered, which follows from the fact that his Z-score is the highest.

Adjustments based on Z-scores are sensitive to the statistics used to calculate the averages and standard deviations used to form the Z-scores. For instance, in the receiving yards example, we used the receiving yards for all players with at least 1 reception in the given year. It could be argued that such a group of players is too large to be directly relevant to the league leaders in receiving yardage and that we should compare the best receivers to the other top receivers in the league rather than to the set of all receivers.

An alternative approach is to calculate the average and standard deviation for some set of top performers each year. Here I used the top $2 \times T$ players in receiving yardage, where $T$ is the number of teams in the league in a given year. For example, for 2012, I used the top 64; for 1960, I used the top 26. Table 3.8 contains the average and standard deviation of this set of receivers, together with the Z-score of the top receiver listed in Table 3.5.

The adjustment procedure is the same as that used previously, except that the averages and standard deviations from Table 3.8 are used instead of those from Table 3.6. The adjusted values are in Table 3.7, under the column "Adj-Top." Note that the ranking of the players is the same for the two adjustments; Johnson's 2012 performance is the best, followed by Harrison in 2002. Rice's 1848 yards in 1995, which ranks second all time behind Johnson, is ranked last among those players' seasons considered. Using

**TABLE 3.8**   Mean and standard deviation of receiving yards for the top players in receiving yards for the years represented in Table 3.5

| YEAR | AVERAGE | SD | Z-SCORE |
|------|---------|-----|---------|
| 2012 | 910.8 | 311.9 | 3.377 |
| 2002 | 918.1 | 258.4 | 3.111 |
| 1995 | 948.2 | 317.5 | 2.834 |
| 1980 | 799.1 | 187.5 | 2.885 |
| 1971 | 616.3 | 167.1 | 2.955 |
| 1960 | 678.6 | 203.8 | 3.039 |

the adjustment based on all players, Johnson is more than 200 yards above any player; using the adjustment based on only the top players, which seems more appropriate in this example, all 6 players are fairly close to one another.

# 3.15 APPLYING PROBABILITY THEORY TO STREAKS IN SPORTS

Probability theory is useful for understanding the role of randomness in sports statistics. In this section, we consider a detailed example of this by looking at how the variability in outcomes naturally leads to "streaks." Streaks, such as consecutive game hitting streaks or consecutive pass completion streaks, always seem to be of great interest in sports. However, an analysis of the probability theory underlying such streaks shows that, even when the outcomes are totally random, relatively long streaks are not uncommon.

We consider the same basic scenario we considered when describing the binomial distribution: We have an experiment and event $A$ of interest. Define a random variable $X$ such that $X = 1$ if $A$ occurs and $X = 0$ otherwise; let $\pi = P(A) = P(X = 1)$. Suppose that we observe a sequence of $n$ independent experiments and let $X_1, X_2 ..., X_n$ be the corresponding values of $X$. Then, $X_1, X_2 ..., X_n$ is a sequence of independent random variables, each taking the value 0 or 1 depending on whether or not $A$ occurs in the experiment. In Section 3.13, we considered $S = X_1 + X_2 + \cdots + X_n$, which has a binomial distribution.

Here, we are concerned with the longest consecutive streak of ones in $X_1, X_2, ..., X_n$, a random variable that we will denote by $L$. The distribution of $L$ will depend on two parameters, $n$, the number of experiments under consideration, and $\pi$, the probability that $A$ occurs in any one experiment. For instance, consider a hitting streak in baseball. We take as the experiment a game in which the batter plays so that $n$ is the number of games played and $\pi$ is the probability of a hit in any given game.

For given values of $n$ and $\pi$, it is possible to calculate the probability distribution of $L$. For instance, suppose that $n = 3$ and $\pi = 1/2$. There are eight possibilities for a sequence of 3 ones and zeros: (0,0,0), (0,0,1), (0,1,0), (0,1,1), (1,0,0), (1,0,1), (1,1,0), (1,1,1). Each of these has probability 1/8. The corresponding values of $L$ are 0, 1, 1, 2, 1, 1, 2, 3, respectively. Therefore, $P(L = 0) = 1/8$, $P(L = 1) = 1/2$, $P(L = 2) = 1/4$, $P(L = 3) = 1/8$. The case for general $n$ and $\pi$ is much more complicated, and this simple method of determining the distribution cannot be used, but the basic idea is the same, and there are published formulas that give the distribution.

Consider the case of a hitting streak. The analysis depends heavily on the player under consideration, so let us look at Miguel Cabrera in the 2011 season. Cabrera played in 161 games, so take $n = 161$. Of those 161 games, Cabrera had at least one hit in 127 games, so let us take $\pi = 127/161 = 0.79$. Using these values, the distribution of $L$, the longest hitting streak of the season, is given in Table 3.9. This distribution gives the probability of a hitting streak of length $L = a$ under the assumptions that the probability that Cabrera gets a hit in a given game is 0.79 and that the game-to-game results are independent.

**TABLE 3.9**  Distribution of the longest hitting streak based on Cabrera's data

| a | P(L = a) |
|---|---|
| <10 | 0.018 |
| 10 | 0.031 |
| 11 | 0.053 |
| 12 | 0.074 |
| 13 | 0.089 |
| 14 | 0.096 |
| 15 | 0.095 |
| 16 | 0.089 |
| 17 | 0.080 |
| 18 | 0.069 |
| 19 | 0.059 |
| 20 | 0.049 |
| 21 | 0.040 |
| 22 | 0.033 |
| 23 | 0.026 |
| 24 | 0.021 |
| 25 | 0.017 |
| 26 | 0.013 |
| 27 | 0.010 |
| 28 | 0.008 |
| 29 | 0.006 |
| 30 | 0.005 |
| >30 | 0.019 |

Based on this distribution, we see that the probability that Cabrera has a hitting streak of at least 20 games is about 0.25 (obtained by adding the probabilities of a streak of length 20 games, 21 games, and so on), so that Cabrera's hitting streak of 17 games in 2011 is not particularly long given his batting ability. In fact, the mean value of $L$ based on $n = 161$ and $\pi = 0.79$ is 16.9, so that Cabrera's 17-game hitting streak is almost exactly what would be expected from a batter with his ability if results in different games are independent. Based on this analysis, we could conclude that there is little evidence of "streakiness" in Cabrera's batting.

The distribution in Table 3.9 is based on Cabrera's statistics. Properties of hitting streaks in general can be derived from some basic assumptions. Take $n = 162$, the length of an MLB season. To determine $\pi$, suppose that the player has 4 at bats per game, and the probability of a hit on any given at bat is $r$; for example, for a .300 hitter, $r = 0.3$. Then, the probability of no hits in 4 at bats is $(1 - r)(1 - r)(1 - r)(1 - r)$; therefore, the probability of at least one hit in a game is

$$\pi = 1 - (1 - r)^4.$$

**TABLE 3.10**   Distributions of streak length

| LENGTH | BATTING AVERAGE | | |
|---|---|---|---|
| | .200 | .250 | .300 |
| <10 | 0.563 | 0.167 | 0.017 |
| 10 to 15 | 0.424 | 0.730 | 0.609 |
| 16 to 20 | 0.012 | 0.087 | 0.269 |
| >20 | 0.001 | 0.015 | 0.105 |
| | | | |
| Mean | 8.6 | 11.4 | 15.0 |

Table 3.10 gives some general properties of the distribution of $L$ for a .200 hitter, a .250 hitter, and a .300 hitter. From this, we see that the probability that a .200 hitter has a batting streak of greater than 20 games in a season is about 0.001; that is, it is extremely unlikely for a .200 hitter to have a hitting streak of longer than 20 games based solely on the random nature of batting. For a .300 hitter, on the other hand, the average longest hitting streak for the season is about 15 games, and a hitting streak of greater than 20 games is not uncommon based only on the randomness of game-to-game results.

Exact calculation of the distribution of the longest streak is difficult and requires specialized software that is not typically available in spreadsheets or statistical packages. However, there is a simple expression for a "typical streak" based on the values of $n$ and $\pi$, given by

$$-\frac{\ln(n(1-\pi))}{\ln(\pi)}.$$

Here, ln represents the natural logarithm function. For instance, for the Cabrera example, where $n = 161$ and $\pi = 0.79$,

$$-\frac{\ln(161(1-0.79))}{\ln(0.79)} = 14.9$$

which is a reasonable value, given the information in Table 3.9.

In addition to usefulness for numerical calculations, this expression for a typical streak is useful for understanding the properties of streaks. Note that the typical value does not depend heavily on the value of $n$ because the ln function changes fairly slowly with $n$ for the values of $n$ typically encountered in sports.

For instance, for a given season, the typical length of a hitting streak for Miguel Cabrera was just shown to be 14.9 games. Now, suppose that we are interested in his longest hitting streak over the past 8 years, assuming that the values of $n$ and $\pi$ that we used for 2011 apply for each year in that entire period. Then, $n = 1288$ and $\pi = 0.79$ and using the preceding equation, the typical length of his longest hitting streak over that period is 23.8 games. That is, although the time period is eight times as long, the typical longest streak is less than twice that of a single year. In fact, during that time period, Cabrera's longest hitting streak was 20 games.

# 3.16  USING PROBABILITY THEORY TO EVALUATE "STATISTICAL ODDITIES"

Sports articles and broadcasts often give much attention to "unusual" results, results that seem to require an inordinate amount of luck or coincidence. In some cases, it is possible to use probability theory to evaluate the likelihood of a particular event occurring. We have seen one instance of this type of analysis, in Section 3.15 on streaks. In this section, we consider three examples of events that might be considered to be unusual and show how to use probability to obtain a rough idea of how likely these events are.

In the May 2, 2011 edition of *Bleacher Report* (http://www.bleacherreport.com), Peter Wardell contributed an entertaining article on "20 Statistical Oddities from the 2011 MLB Season So Far." Here we look at three of these "oddities": the fact that 2 teams had a sub-50% save conversion rate; the fact that the Nationals had only 1 pinch hit in 26 pinch-hit at bats; and the fact that Kurt Suzuki had a 55% caught-stealing percentage.

First, consider the save conversion rates. After the first month of the 2011 MLB season, the White Sox had a save conversion rate of 33%, and the Astros had a save conversion rate of 36%. As noted in the article, in the previous decade only two teams had a season save conversion rate of less than 50% (and both were close, at 48% and 49%).

The save conversion rate is the ratio of saves to the sum of saves plus blown saves, expressed as a percentage. For convenience, I refer to saves plus blown saves as "save attempts." In a month of an MLB season, teams average about 10 save attempts; the average save conversion rate of MLB teams is about 68%. Let $X$ denote the number of save conversions for a team. Assuming the probability of a save conversion is 0.68 and that a team has 10 save attempts, we can model $X$ as a random variable with a binomial distribution with $n = 10$ and $\pi = 0.68$. The probability that the team has a save conversion rate of less than 50% is $P(X \leq 4)$. Note that if $X$ is less than or equal to 4, the save conversion rate is 40% or lower.

Using the probability function of the binomial distribution, it can be shown that

$$P(X \leq 4) = 0.064.$$

Therefore, it appears that having two teams with such a low save conversion rate is unlikely. However, it is important to keep in mind that there are 30 MLB teams, each of which might have a low save conversion rate.

Because the probability of a "low" save conversion rate (i.e., one that is less than 50%) is 0.064, the probability of a "normal" save conversion rate (i.e., one that is 50% or greater) is

$$1 - 0.064 = 0.936.$$

The chance that all 30 MLB teams have a normal save conversion rate, assuming that they all have 10 save attempts, the probability of a save conversion is 0.68 for all teams, and that the results of different teams are independent, is

$$(0.936)^{30} = 0.139.$$

Using properties of the binomial distribution, it may be shown that the probability that all but 1 team has a normal save conversion rate is 0.283. Therefore, the probability that 2 or more teams have a low save conversion rate is

$$1 - 0.139 - 0.283 = 0.578.$$

That is, it is actually more likely that at least 2 teams will have a low save conversion rate in the first month of the season than it is that all teams, or all but 1 team, will have a normal save conversion rate in the first month. Although the probability 0.578 is based on a number of assumptions that are unlikely to be exactly true, it seems reasonable to conclude that the fact that 2 teams had a sub-50% save conversion rate in the first month of the season is not particularly unusual.

Now, consider the fact that the Nationals were 1 for 26 in the first month of the 2011 season. Because only National League teams have a large number of pinch-hit at bats, the analysis here refers only to National League teams.

The batting average of pinch hitters for the league overall was .214 in 2011. Let $X$ denote the number of pinch hits a team has in 26 pinch-hit at bats; then, we can model $X$ as a binomial random variable with $n = 26$ and $\pi = 0.214$. It follows that the probability that a team has 0 hits in 26 at bats is

$$P(X = 0) = (1 - 0.214)^{26} = 0.0019;$$

the probability that the team has 1 hit in 26 at bats is 0.0135. Therefore, the probability that a team has less than or equal to 1 hit in 26 pinch-hit at bats is 0.0154; it follows that the event that a team goes 1 for 26 in pinch hitting is unusual.

However, as in the save conversion example, we must take into account the fact that there are 16 National League teams, each of which had a chance to go 1 for 26 in pinch hitting. Because the probability that a given team has at least 2 hits in 26 at bats is $1 - 0.0154 = 0.9846$, the probability that all 16 teams have at least 2 hits in 26 at bats is

$$(0.9846)^{16} = 0.780.$$

Stated another way, the probability that at least 1 team has 0 or 1 hit in 26 pinch-hit at bats is 0.22. We expect this to occur about once every $1/0.22 = 4.5$ seasons. Therefore, according to this analysis, the fact that a team went 1 for 26 in pinch hitting to start the season is fairly unusual. Furthermore, part of the "oddity" of this result, as described by Wardell, is that the Nationals had 9 strikeouts in 26 at bats, making their first-month pinch-hitting performance even more unusual.

Finally, consider the fact that Kurt Suzuki threw out 16 of 29 players attempting to steal in the first month of the 2011 season. What makes this event unusual is the fact that Suzuki had a caught-stealing percentage of only 22% in 2010 and 25% in 2009. Therefore, this oddity can be interpreted as one in which a catcher with a relatively poor record of throwing out those attempting to steal has a stretch of 29 attempts in which he throws out 16 (or more).

Of the 30 catchers in 2010 with the most playing time, 15 had a caught-stealing percentage of less than 30%. The average caught-stealing percentage of these 15 catchers was 22.4%, which coincidently is the same as Suzuki's caught-stealing percentage in 2010. Let $X$ denote the number of runners caught stealing by a given catcher; we can model $X$ as a random variable with a binomial distribution with $n = 29$ and $\pi = 0.224$. Under this assumption,

$$P(X \geq 16) = 0.000128.$$

Therefore, the probability that 1 of these 15 catchers throws out no more than 15 of 29 runners is

$$1 - 0.000128 = 0.999872$$

and the probability that all 15 catchers throw out 15 or fewer runners is

$$(0.999872)^{15} = 0.99808.$$

It follows that the probability that a catcher with a poor record of throwing out those attempting to steal throws out 16 (or more) of 29 runners attempting to steal is

$$1 - 0.99808 = 0.00192;$$

that is, this is an extremely rare event that, according to the analysis in this section, can be expected to occur only once every $1/0.00192 = 521$ seasons.

Although, given the number of assumptions used in this analysis, we should not take the 521 seasons result too seriously; it is clear that Suzuki's start to the 2011 season deserves to be called a statistical oddity. It is worth noting that for the remainder of 2011, Suzuki threw out only 22 of 107 runners attempting to steal, and he ended the season with a caught-stealing percentage of about 28%.

## 3.17 COMPUTATION

Probability calculations for the two distributions considered in this chapter, the binomial and the normal distributions, can be easily carried out in Excel.

First, consider the binomial distribution. In Section 3.16, when analyzing the save conversion oddity, it was noted that if $X$ is a random variable with a binomial distribution with $n = 10$ and $\pi = 0.68$, then

$$P(X \leq 4) = 0.064.$$

In Excel, this probability can be calculated using the function BINOM.DIST. For a random variable $Y$ that has a binomial distribution with parameters $n$ and $\pi$, $P(Y \leq a)$ can be obtained from

BINOM.DIST(a, n, $\pi$, TRUE);

the "TRUE" in the statement refers to the fact that we want the cumulative probability $P(Y \leq y)$ rather than the individual probability $P(Y = y)$, which would be calculated using

BINOM.DIST(a, n, $\pi$, FALSE).

Therefore, in the save conversion example, $P(X \leq 4)$ can be obtained using

BINOM.DIST(4, 10, 0.68, TRUE),

which returns the value 0.063715.

Now, consider the normal distribution. Let $X$ denote a random variable with a normal distribution with mean $\mu$ and standard deviation $\sigma$. Then, $P(X < a)$ can be obtained using the Excel command

NORM.DIST(a, $\mu$, $\sigma$, TRUE).

A probability of the form $P(-a < X < a)$ can be expressed as

$$P(X < a) - P(X < -a).$$

Therefore, it can be calculated using

NORM.DIST(a, $\mu$, $\sigma$, TRUE) $-$ NORM.DIST($-a$, $\mu$, $\sigma$, TRUE).

For instance, Table 3.4 gives $P(-1 < Z < 1)$ where $Z$ is a standard normal random variable, that is, a random variable with a normal distribution with mean 0 and standard deviation 1. This probability can be calculated using

NORM.DIST(1, 0, 1, TRUE) $-$ NORM.DIST($-1$, 0, 1, TRUE),

which returns the value 0.682689.

# 3.18 SUGGESTIONS FOR FURTHER READING

Probability theory is an important area of mathematics with applications in a wide range of fields. There are two aspects of probability: the technical side, which focuses on the mathematical properties of probability functions and random variables, and the intuitive side, which focuses on understanding randomness and the role it plays in many fields of study, including sports. For further reading, the works of Grinstead and Snell (1997) and Ross (2006) are detailed introductions to the mathematics of probability, suitable for readers with strong math backgrounds and a desire to understand the technical details behind probability theory. Mlodinow (2008) does an excellent job of describing the intuition behind probability theory and randomness; this book is suitable for a general audience and does not require a background in mathematics.

Win probabilities and expected points are important general techniques used in many sports. See the works of Tango, Lichtman, and Dolphin (2007, Chapter 1); Click (2006); and Woolner (2006) for applications in baseball; for applications in football, see Winston (2009, Chapters 21 and 24), Goldner (2012), and the Advanced Football Analytics website (http://www.advancedfootballanalytics.com).

The method of adjustment described in Section 3.11 is known as poststratification; see the work of Wainer (1989) for a detailed discussion of the pros and cons of this approach, as well as some alternative methods. Poststratification can also be used to adjust for a continuous variable by grouping the variable into classes, similar to the method used in Section 3.12 (although in that case, the continuous variable—the length of a field goal attempt—is not readily available); see Cochran's (1968) work.

The Z-score approach to comparing players from different eras presented in Section 3.14 was used by Lependorf (2012) to compare MLB players from different eras and by Silver (2006a) in his comparison of Babe Ruth and Barry Bonds. Lederer (2009) uses Z-scores in his method for ranking MLB pitchers.

Streaks are a popular topic for discussion in sports. Gilovich (1991, Chapter 2) discusses the tendency to try to explain streaks in terms of some underlying pattern or theory, rather than as simply random occurrences. Moskowitz and Wertheim (2011, pp. 215–229) and Winston (2009, Chapter 11) discuss the properties of streaks in the context of sports. The technical results on streaks in Section 3.15 are based on Schilling's (1990) work.

# Statistical Methods

# 4

---

## 4.1 INTRODUCTION

In an ideal world, we would have an unlimited amount of data, and all relevant questions could be answered with certainty. Is Tom Brady better than Peyton Manning? Have them play hundreds and hundreds of games with the same teammates against similar opponents and analyze the results. Of course, in the real world, this is not possible, and we have to base our analyses on the available data.

Statistical methods play at least two roles in these situations. First, they provide methods for extracting the maximum amount of information from a set of data. Second, they give us a way to quantify the uncertainty that results from having to base these conclusions on such limited data.

The goal of this chapter is to give an overview of statistical reasoning and the type of statistical methods that are useful in analyzing sports data.

---

## 4.2 USING THE MARGIN OF ERROR TO QUANTIFY THE VARIATION IN SPORTS STATISTICS

One goal of analytic methods is to use data to draw conclusions about the process generating the data. For instance, suppose we want to study the performance of an NFL (National Football League) running back. Let $Y$ denote the yards gained on a particular carry and let $Y_1, \ldots, Y_n$ denote the results of the running back's carries in a given season.

Then, $Y$ can be modeled as a random variable; its distribution is a way to describe the running back's "true ability level," which plays a central role in his performance. This true ability level is unknown, but its properties are reflected in the observed data. A simple model for relating the observed result to this hypothetical true ability level is to assume that $Y_1, \ldots, Y_n$ are independent random variables each with the distribution of $Y$. More complicated models might take into account the fact that different carries occur in different situations (e.g., down, distance to first down) and against different defenses. However, for now, the focus is on the simple model.

Under this simple model, the probability distribution of $Y$ completely characterizes the running back's ability on running plays. Therefore, if we know this distribution, we can use it to help predict future results, to compare this player to other running backs, to aid in play calling, and so on. However, because the distribution is unknown, we must use the information in $Y_1, Y_2, \ldots, Y_n$ to learn about the probability distribution of $Y$. The practice of using sample data to determine the properties of the underlying probability distributions is known as statistical estimation or simply as estimation.

In many cases, estimation can be based on the analogy principle. To estimate a characteristic of the distribution of $Y$, we use the corresponding characteristic of the data $Y_1, Y_2, \ldots, Y_n$. For instance, to estimate the mean of $Y$, we use the sample mean of $Y_1, Y_2, \ldots, Y_n$; to estimate the standard deviation of $Y$, we use the sample deviation of $Y_1, Y_2, \ldots, Y_n$, and so on.

Suppose the running back in question is Jamaal Charles in the 2010 NFL season. Using the data described in Section 2.3, we can estimate properties of the distribution of $Y$. For instance, the sample mean of the observed $Y$ values is 6.38 yards per carry; therefore, our estimate of the mean of $Y$ is 6.38 yards per carry. Similarly, the sample median of the observed $Y$ values is 4 yards per carry, yielding an estimate of the median of $Y$.

When interpreting results of this type, it is important to recognize the inherent variability in sports results. For instance, if we were somehow able to replay the 2010 NFL season, we would not expect Charles to have a yards-per-carry value of exactly 6.38 again, although we would expect it to be "close to" 6.38. Furthermore, because of this variability, we know that the observed statistics do not exactly measure a player's true ability; for instance, we do not expect Charles's "true yards per carry" for 2010 to be exactly 6.38 yards, although, again, we expect that it will be close to 6.38 yards. Here we can think of his true yards per carry as the yards per carry that would be obtained if we were able to observe a large number of rushing attempts by Charles; alternatively, we can think of it as the mean of the random variable $Y$ representing the results of a Charles rushing attempt.

Of course, simply saying that if we were to replay the 2010 NFL season, Charles's yards per carry would be close to 6.38 yards or his true yards per carry is close to 6.38 yards is not enough—we need a quantitative measure of what *close to* means. In both cases, this is given by a statistical measure known as the *margin of error*.

To understand what the margin of error is measuring, consider the following experiment: Suppose we simulate Charles's performance during the 2010 NFL season by randomly choosing the results of 230 rushing attempts; recall that in the actual 2010 NFL season, Charles had 230 rushing attempts. An important consideration in carrying out this experiment is the probability distribution to use for the simulation. The best information we have about the distribution of the yards gained on Charles's rushing attempts is the actual results. Therefore, to simulate the result of his first rushing attempt, we randomly select one result from his 230 actual results; to simulate the result of his second rushing attempt, we select another result from his 230 actual results, and so on. Doing this 230 times gives us a simulated season for Charles; when I carried out such a process, I observed a simulated yards per carry of 6.54 yards.

The difference between the simulated value of 6.54 and the actual value of 6.38 gives us important information regarding the natural variability in Charles's yards per carry. However, we could obtain more information by simulating the 2010 season several times; I did this, obtaining yards-per-carry values of 6.37, 6.29, 6.15, 5.71, 7.21, and so on. The margin of error can be viewed as a measure of the variability of a large number of such simulated values; specifically, it is two times the standard deviation of a long sequence of simulated values. In the example of Charles's yards per carry, it is 1.16 yards.

We can interpret the margin of error in one of two ways. The most straightforward interpretation is based on the idea of repeating the "experiment" (e.g., a season, a game, or a career) and calculating a range of values that we expect to include the statistic of interest. For instance, in Charles's rushing example, if we were to repeat the 2010 NFL season, we would expect Charles's yards-per-carry value to fall in the range

$$6.38 \pm 1.16 = (5.22, 7.54).$$

Therefore, the interval (5.22, 7.54) summarizes the variability in Charles's yards-per-carry value.

A second interpretation, based on the underlying "true" characteristic of a player or team, is a little more subtle, but it is often more useful. Recall that a statistical estimate is our "best guess" of the characteristic of a probability distribution. However, we know that the estimate is not equal to that characteristic. For example, in the example for Charles, his observed average yards per carry is 6.38 yards; as noted, we expect this to be close to the true mean value of $Y$, but it is unlikely that it will be exactly equal to it. The observed value of 6.38 is an estimate of this true mean value.

The margin of error can also be interpreted in terms of how close we expect this hypothetical true mean value to be to the observed value of 6.38. The interval

$$6.38 \pm 1.16 = (5.22, 7.54)$$

gives a range of values such that we are "reasonably certain" that Charles's true average yards per carry lies in that range.

At this point, it is reasonable to ask why we are interested in the true mean yards per carry or in a range of results that might occur if the season is repeated. In fact, for some purposes we are not particularly concerned with these hypothetical values. This might be the case, for example, if we are just trying to summarize a player's or team's performance.

However, in other cases, we are primarily interested in understanding the process that generated the data to compare players or teams or to better understand what might happen in the future. In these cases, the underlying true values that are of interest and the variability of estimates are an important consideration both in understanding how future results might relate to the available data and in determining how much confidence we should place in the conclusions of the analysis. Either interpretation of the margin of error gives us important information about the variability of estimates and their relationship to the hypothetical true value.

For example, suppose we are analyzing Kevin Durant's scoring for the 2011–2012 NBA (National Basketball Association) season. Let $X$ denote Durant's points scored for a given game, and suppose we model $X$ as a random variable with some probability distribution. Although different aspects of this distribution might be of interest, depending on the context, here we focus on the mean of $X$, which we note by $\mu$. The available data are the points scored for the 66 games Durant played, and using the analogy principle, our estimate of $\mu$ is simply the average of the 66 values, 28.0, Durant's points per game (PPG) for the 2011–2012 regular season.

Although 28.0 exactly represents what actually occurred in the 2011–2012 season, it is only an estimate of $\mu$, Durant's "true PPG," that is, the PPG he would achieve in a hypothetical long sequence of games. The uncertainty in this estimate is described by the margin of error. Here, the margin of error is 1.7; how this was determined is discussed in the following section. Therefore, our estimate of $\mu$, Durant's true PPG for 2011–2012, could be reported as

$$28.0 \pm 1.7$$

or as the interval (26.3, 29.7). That is, if we were somehow able to observe a long sequence of Durant's games, under the conditions of the 2011–2012 season, our best guess for his average PPG in this sequence is 28.0, and we are reasonably certain that it would fall between 26.3 and 29.7.

Alternatively, we can use the interpretation of the margin of error in terms of repeating the season. That is, if we were able to repeat the 2011–2012 NBA season, we would expect Durant's PPG value to lie in the range 26.3 to 29.7.

There are two basic approaches to calculating the margin of error. The first is to use statistical formulas that are designed for this purpose and are included as part of many spreadsheet and statistical packages. This approach is discussed in Section 4.3. The second approach is based on the interpretation of the margin of error in terms of hypothetical repetitions of the experiment. That is, we can compute the margin of error using computer simulation to obtain these hypothetical repetitions; this approach is discussed in Section 4.4.

# 4.3 CALCULATING THE MARGIN OF ERROR OF AVERAGES AND RELATED STATISTICS

Although we will not generally calculate the margin of error by hand, using statistical formulas, it is useful to consider the margin-of-error formula in a few simple cases to better understand the issues that drive the accuracy of an estimate.

Consider the Durant scoring example from the previous section. Let $X$ denote Durant's points scored for a given game and suppose we are interested in $\mu$, the mean of $X$. Our estimate of $\mu$ is the sample mean of Durant's points scored for the 66 games he played. The formula for the margin of error in this case, in which we are estimating the mean of a random variable using a sample mean, is $2S/\sqrt{n}$, where $S$ denotes the sample standard deviation of his game-by-game points scored and $n$ denotes the number of data values used in the sample average. For this example, $n = 66$, and $S$ can be calculated by examining his game-by-game statistics, which show that $S = 6.9$. Thus, the margin of error is

$$2\frac{S}{\sqrt{n}} = \frac{2(6.9)}{\sqrt{66}} = 1.7$$

as reported in the previous section.

The formula $2S/\sqrt{n}$ for the margin of error in this case shows that the accuracy of the sample average as an estimate of the mean of a random variable is driven by two factors. One is the amount of information available for estimation, as reflected in the sample size $n$: The larger the sample size, the smaller the margin of error is. However, note that doubling the sample size does not cut the margin of error in half; because it is $\sqrt{n}$ in the formula, not $n$, doubling the sample size corresponds to dividing the margin of error by about 1.4.

The other factor affecting the margin of error is the natural variability in the data, as measured by the standard deviation $S$. Therefore, measurements that are approximately constant have a small margin of error for estimating the mean, while measurements that show a lot of variability have a large margin of error. Table 4.1 contains the margins of error for several statistics based on Durant's 2011–2012 season (Dataset 4.1).

Although the formula for the margin of error of an average is based on the assumption that we have a sequence of observations with the same mean and standard deviation, the result is still useful even if the means and standard deviations of the random variables vary, provided that we are interested in the overall average of the observations. In that case, $2S/\sqrt{n}$ tends to slightly overestimate the variability in the sample mean, but this overestimation is generally small.

For instance, in the Durant example, it may be appropriate to assume that the distribution of his points scored depends on whether the game is played at home or away, and in fact, his average PPG value is higher for home games, 28.8 PPG, compared to 27.2 for away games. Under this assumption, the margin of error for his overall scoring average (based on more sophisticated methods beyond the scope of this book) is 1.695; the margin of error using $2S/\sqrt{n}$ is 1.693. Hence, in this case, the margin of error based on $2S/\sqrt{n}$ is slightly less than the correct margin of error; however, the difference is negligible.

**TABLE 4.1**   Results from Durant's 2011–2012 season

| STATISTIC | MEAN | SD | MARGIN OF ERROR |
|-----------|------|-----|-----------------|
| Rebounds  | 7.98 | 3.03 | 0.74 |
| Assists   | 3.50 | 1.95 | 0.48 |
| Turnovers | 3.76 | 1.70 | 0.42 |
| Fouls     | 2.02 | 1.36 | 0.34 |
| Points    | 28.03 | 6.88 | 1.69 |

Another commonly used statistic with a simple formula for the margin of error is a proportion. Consider an experiment and let $A$ denote an event of interest; let $\pi = P(A)$ be the probability of $A$ occurring. Suppose that the experiment is performed $n$ times and let $X$ denote the number of times that $A$ occurs. We have seen that $X$ has a binomial distribution with parameters $n$ and $\pi$. We can use $X$ to estimate $\pi$. Define $p = X/n$; then, $p$ is the proportion of experiments in which event $A$ occurs. That is, $p$ is the sample analogue of $\pi$, and it can be used as an estimator of $\pi$. The formula for the margin of error of this estimator is

$$2\,\frac{\sqrt{p(1-p)}}{\sqrt{n}}.$$

Note that this expression has the same general form as the one for the margin of error of an average: The denominator is the amount of information as reflected in the sample size, and the numerator is a measure of the variability of the data. Because the data are particularly simple—for each experiment, either $A$ occurs or it does not occur—the measure of variability is a function of the parameter estimate being analyzed.

Consider the following example: In 2012, Buster Posey had an on-base average (OBA) of 0.408 based on 610 plate appearances. The margin of error corresponding to his OBA is therefore

$$\frac{2\sqrt{0.408(1-0.408)}}{\sqrt{610}} = 0.0398$$

so that we are reasonably certain that Posey's hypothetical true OBA for 2012 is in the interval (0.368, 0.448).

Note that, even with a sample size of 610 plate appearances, the margin of error is relatively large. This is often the case when estimating proportions. Table 4.2 gives the margins of error, based on an observed proportion of 0.400, for a wide range of sample sizes. Many statistics, such as batting average and OBA are traditionally reported to three decimal places. The results in Table 4.2 show that, even for a large sample size of 10,000 (roughly 15 seasons of at bats, e.g.), the third digit in such a statistic is not particularly meaningful; this conclusion follows from the fact that the margin of error in that case is 0.005.

TABLE 4.2    Margins of error based on a proportion of 0.400

| SAMPLE SIZE | MARGIN OF ERROR |
| --- | --- |
| 500 | 0.022 |
| 1,000 | 0.015 |
| 2,000 | 0.011 |
| 5,000 | 0.007 |
| 10,000 | 0.005 |

Note that if either $X = 0$ or $X = n$, so that either $p = 0$ or $p = 1$, then the margin of error is 0. Although analyses in which an event either never occurs or always occurs are not common, this sometimes occurs when the sample size is very small. Because in these cases we do not expect that the true probability is either 0 or 1 (if we did, it is unlikely that we would be doing the analysis), using 0 as the margin of error is misleading. In these cases, a simple solution is to change $X$ to $X + 1/2$ and to change $n$ to $n + 1$ when calculating $p$ to use in the margin of error. For instance, if our event of interest never occurs in $n = 9$ trials, so that $X = 0$, we use

$$\frac{1/2}{9+1} = 0.05$$

to calculate the margin of error, leading to a margin of error of

$$2\frac{\sqrt{(0.05)(1-0.05)}}{\sqrt{9}} = 0.145;$$

in this case, we can interpret the margin of error as one directional, in the sense that 0 to 0.145 is a range of reasonable values for the true underlying probability of the event.

As with the case of the margin of error of an average, the formula for the margin of error of a proportion can also be used if the underlying probability of the event varies from trial to trial, provided that we are interested in estimating the overall average probability. For instance, in the Posey example, we might think that his on-base percentage (OBP) depends on whether the pitcher is right- or left-handed. This is supported by his 2012 results; his OBP against right-handed pitchers is 0.382, compared to 0.470 against left-handed pitchers. Using a formula for the margin of error of his overall OBP that takes into account this difference (using methods beyond the scope of this book), the margin of error is 0.0396. Recall that the margin of error we calculated previously is 0.398, slightly higher but still very close to the correct value.

This discussion of the margin of error of a proportion applies when there are two outcomes of interest, such as "on base" and "not on base" for a given plate appearance. A similar formula applies when there are several possible outcomes. Suppose that for a given observation one of $m$ events $A_1, A_2, \ldots, A_m$ occurs, and suppose that we are interested in a statistic that assigns weight $w_j$ to event $A_j$. For instance, consider the slugging average of a MLB (Major League Baseball) player, in which the events of interest are "single," "double," "triple," "home run," and "no hit." Slugging average assigns weight 1 to single, 2 to double, 3 to triple, 4 to home run, and 0 to no hit.

For a given set of data, based on $n$ observations, let $p_j$ denote the proportion of experiments in which $A_j$ occurs. Then, the average value of the statistic for the dataset is given by

$$w_1 p_1 + w_2 p_2 + \cdots + w_m p_m.$$

For example, consider Mike Trout in 2012. In his 559 at bats, he had 117 singles, 27 doubles, 8 triples, and 30 home runs. Therefore, if $A_1$ represents a single, $p_1 = 117/559 = 0.2093$; similarly, letting $A_2$ denote a double, $p_2 = 0.0483$; letting $A_3$ denote a triple, $p_3 = 0.0143$; and letting $A_4$ denote a home run, $p_4 = 0.0537$. It follows that his slugging average is

$$0.2093 + 2(0.0483) + 3(0.0143) + 4(0.0537) = 0.564.$$

Note that because no hit receives weight 0 in the slugging percentage calculation, we did not need to include $p_5$, the proportion of at bats in which Trout did not get a hit, in the calculation.

For a statistic of the form $T = w_1 p_1 + w_2 p_2 + \cdots + w_m p_m$, the formula for the margin of error is

$$2 \; \frac{\sqrt{(w_1^2 p_1 + w_2^2 p_2 + \cdots + w_m^2 p_m) - T^2}}{\sqrt{n}}.$$

For example, for Trout, $T = 0.564$,

$$w_1^2 p_1 + \cdots + w_m^2 p_m = 1(0.2093) + 4(0.0483) + 9\,(0.0143) + 16\,(0.0537) = 1.3904$$

and $n = 559$, so that the margin of error is

$$2 \; \frac{\sqrt{1.390 - (0.564)^2}}{\sqrt{559}} = 0.088.$$

It follows that Trout's slugging percentage can be described as $0.564 \pm 0.088$ or as the interval $(0.476, 0.652)$.

Note the margin of error for Trout's slugging percentage is much larger than the margin of error for a batting average of 0.564 based on 559 at bats, which would be

$$2 \; \frac{\sqrt{0.564\,(1 - 0.564)}}{\sqrt{559}} = 0.042.$$

This is because of the weighting that is used when calculating slugging percentage: By weighting home runs by 4, triples by 3, and doubles by 2, the natural randomness in the events home run, triple, and double is magnified. On the other hand, for batting average, every event receives either a weight of 1 (hits) or a weight of 0 (nonhits). This is generally true; statistics that are calculated by giving some events relatively large weight tend to have large margins of error. That is, there is often more uncertainty associated with the statistic than we might normally expect.

# 4.4 USING SIMULATION TO MEASURE THE VARIATION IN MORE COMPLICATED STATISTICS

In Section 4.2, an interpretation of the margin of error was given in terms of hypothetical repetitions of the "experiment," such as a game or season, under consideration. The margin of error reflects the fact that if data are sampled from a given probability distribution by, for example, observing a player or team over the course of a season, the actual results observed will reflect the natural randomness of the process generating the data. In this section, we consider calculation of the margin of error by simulating this process. This approach is useful for more complicated statistics, for which the margin of error cannot be obtained from Excel or from statistical software.

We begin by considering the example of Durant's per game scoring in 2011–2012. Although this statistic can be handled using the techniques of the previous section, it is instructive first to apply the simulation-based method to a simple case before considering more complicated ones.

Let $X$ denote his points scored in a given game and, based on data from the 2011–2012 season, the average value of $X$ is 28.0. The margin of error of this estimate measures the natural variability in this estimate. That is, if we were somehow able to replay the 2011–2012 NBA season, Durant's game-by-game points scored values, as well as his season average PPG would be different, reflecting this natural variability.

To understand the variability in his average PPG, we can "simulate" a 2011–2012 NBA season, or at least Durant's scoring for that season. To do this, we can simulate 66 points-scored values from a probability distribution that represents Durant's scoring. Based on these values, we can compute an average PPG value for this simulated season. If we repeat this process many times, we obtain simulated average PPG values $\bar{X}_1, \bar{X}_2, \ldots, \bar{X}_M$, where $M$ denotes the number of simulated seasons. The variation in these simulated averages gives us some information about the variability of Durant's average PPG; specifically, the margin of error is given by $2\hat{S}$, where $\hat{S}$ denotes the sample standard deviation of $\bar{X}_1, \bar{X}_2, \ldots, \bar{X}_M$.

The remaining issue is the choice of the probability distribution to use in the simulation. We want that distribution to represent Durant's PPG, but that distribution is, of course, unknown. The best information we have about Durant's PPG is the actual values from the 2011–2012 season, and we can use those values as the basis for the simulation.

The 66 PPG values from Durant's 2011–2012 season are given in Table 4.3. To sample from this distribution, we simply randomly choose 66 new values from the values in Table 4.3. To do this, we could choose a random integer from 1 to 66 and choose the corresponding value from Table 4.3. For instance, if the random number is 6, Durant's points scored for the first game in his simulated season is 24; if the second random number drawn is 56, Durant's points scored for the second game in his simulated season is 22, and so on.

**TABLE 4.3**    Durant's game-by-game scoring in 2011–2012

| | | | | | | | | | | | | | | | | | |
|---|---|---|---|---|---|---|---|---|---|---|---|---|---|---|---|---|---|
| 32 | 32 | 35 | 29 | 29 | 24 | 43 | 29 | 22 | 19 | 23 | 44 | 30 | 21 | 26 | 21 | 25 | 28 |
| 40 | 32 | 18 | 26 | 25 | 24 | 28 | 26 | 23 | 30 | 22 | 35 | 38 | 23 | 33 | 28 | 31 | 51 |
| 23 | 33 | 21 | 19 | 27 | 33 | 33 | 22 | 36 | 23 | 36 | 37 | 25 | 20 | 20 | 33 | 28 | 28 |
| 29 | 22 | 21 | 27 | 26 | 19 | 27 | 12 | 30 | 32 | 33 | 30 | | | | | | |

Suppose we repeat this process until we have 66 values, corresponding to an entire simulated 2011–2012 season. An example of such a dataset is given in Table 4.4. The average of the 66 values in Table 4.4 is 27.7. As expected, this is close to Durant's actual scoring average for 2011–2012, but it is not exactly the same; the difference between the values gives us some indication of the variability in scoring averages over a 66 game season.

More information about this variability can be obtained by simulating several 2011–2012 seasons for Durant. As discussed previously, let $\bar{X}_1, \bar{X}_2, \ldots, \bar{X}_M$ denote the simulated scoring averages for $M$ simulated seasons; therefore, $\bar{X}_1 = 27.7$. I repeated the entire procedure nine times so that we have the results of 10 simulated seasons; Durant's simulated scoring averages for these seasons are 27.7, 28.8, 27.9, 26.9, 27.0, 27.1, 29.6, 28.7, 28.9, 28.5. The mean of these values is 28.1, which is close to Durant's actual average PPG for the 2011–2012 season. However, it is the variability of these values that is of main interest to us.

These values show how we might expect the average PPG of a player with Durant's skill level, for a 66-game season, to vary. The values are all relatively close to Durant's actual average PPG of 28.0, but they range from 26.9 to 29.6. The standard deviation of these values is 0.92, and twice this standard deviation, 1.84, can be used as the margin of error for his average PPG. This is close to, but not exactly equal to, the value we obtained using the margin-of-error formula (1.69). The advantage of the simulation-based approach is that we do not need to know the formula to calculate the margin of error.

Of course, it is important to recognize that the value 1.84 is based on the random numbers used in the simulation. I repeated the entire process, simulating 10 more seasons and computing the margin-of-error based on the results; this procedure yielded the margin of error 1.55, again close to the formula-based result but not exactly equal to it.

It is natural to ask why I am only using 10 seasons: Would more seasons lead to a better estimate of the margin of error? The answer is yes, more seasons, that is, a larger value of $M$, would be preferable. Table 4.5 contains the margins of error based on several different values of $M$; for each value of $M$, the entire simulation was repeated 1000 times, and the median, the upper and lower quartiles, and the

**TABLE 4.4**    Simulated game-by-game scoring by Durant

| | | | | | | | | | | | | | | | | | |
|---|---|---|---|---|---|---|---|---|---|---|---|---|---|---|---|---|---|
| 24 | 22 | 27 | 24 | 23 | 26 | 22 | 38 | 28 | 40 | 25 | 28 | 37 | 35 | 35 | 29 | 22 | 24 |
| 19 | 20 | 22 | 28 | 19 | 25 | 26 | 32 | 25 | 21 | 22 | 30 | 28 | 26 | 36 | 26 | 33 | 21 |
| 33 | 35 | 26 | 32 | 29 | 21 | 35 | 21 | 28 | 19 | 19 | 40 | 33 | 28 | 26 | 27 | 40 | 38 |
| 27 | 22 | 32 | 22 | 33 | 29 | 19 | 38 | 25 | 29 | 22 | 33 | | | | | | |

**TABLE 4.5**   Distribution of simulated margins of error for different simulation sizes

| M | MINIMUM | LOWER QUARTILE | MEDIAN | UPPER QUARTILE | MAXIMUM |
|---|---|---|---|---|---|
| 10 | 0.62 | 1.37 | 1.63 | 1.88 | 3.10 |
| 25 | 0.91 | 1.51 | 1.67 | 1.83 | 2.34 |
| 50 | 1.17 | 1.56 | 1.68 | 1.79 | 2.20 |
| 100 | 1.35 | 1.60 | 1.68 | 1.76 | 2.05 |
| 500 | 1.51 | 1.64 | 1.68 | 1.72 | 1.86 |
| 1,000 | 1.54 | 1.66 | 1.68 | 1.71 | 1.79 |
| 10,000 | 1.63 | 1.67 | 1.68 | 1.69 | 1.72 |

minimum and maximum are reported, showing the values that typically occur. In practice, the analyst would only do the simulation once; the purpose of showing the results from 1000 simulations is to show the effect of the random numbers used. For instance, suppose we use $M = 100$; there is about a 50% chance of obtaining a margin of error between 1.60 and 1.76, and we are virtually certain to obtain one between 1.35 and 2.05. Note that as the value of $M$ increases, the values in the table approach the formula-based value of 1.69; for example, for $M = 10,000$, even the minimum and maximum values are close to 1.69.

The conclusion here is that, all things being equal, it is better to use a larger value of $M$. However, if we just want a rough idea of the magnitude of the margin of error, using $M = 25$, or even $M = 10$, still gives useful information and is greatly preferable to just ignoring the margin of error completely.

Of course, to determine the margin of error of an average we do not need to use simulation; we can simply use the formula $2S/\sqrt{n}$. The real power in the simulation method is in complicated situations for which there is no formula available. This is generally the case for statistics that are not simple averages or proportions, as in the following section.

# 4.5 THE MARGIN OF ERROR OF THE NFL PASSER RATING

Consider the NFL passer rating. The passer rating is based on four statistics: completions per passing attempt $COMP$, passing yards per attempt $YDS$, touchdown passes per attempt $TD$, and interceptions per attempt $INT$. It is constructed by first standardizing each of these four statistics to put them on roughly equal scales. Therefore, $COMP$ is converted to

$$C = 5\,(COMP - 0.3),$$

*YDS* is converted to

$$Y = 0.25(YDS - 3),$$

*TD* is converted to

$$T = 20\ (TD),$$

and *INT* is converted to

$$I = 2.375 - 25(INT).$$

However, to make sure that no one statistic has an overwhelming influence on the rating, each of *C*, *Y*, *T*, and *I* is truncated to the range (0, 2.375). That is, if a player has a yards-per-completion value of 2, so that

$$Y = 0.25\ (YDS - 3) = 0.25(2 - 3) = -0.25,$$

*Y* = 0 is used instead, and if a player has 20 touchdown passes in 100 attempts, so that *TD* = 0.20 and *T* = (20)(0.20) = 4, *T* = 2.375 is used instead. Finally, the truncated values *C*, *Y*, *T*, and *I* are combined into the overall measure

$$R = 100(C + Y + T + I)/6,$$

which is the NFL passer rating.

For instance, consider Joe Flacco in the 2012 NFL season. Flacco had 317 completions on 531 passing attempts for 3817 yards and 22 touchdowns with 10 interceptions. Therefore, using the notation described, *COMP* = 317/531 = 0.597, *YDS* = 3817/531 = 7.19, *TD* = 22/531 = 0.041, and *INT* = 10/531 = 0.019. It follows that *C* = 1.49, *Y* = 1.05, *T* = 0.82, and *I* = 1.90, leading to a passer rating of 87.7.

We can view Flacco's 2012 passer rating as an estimate of his underlying true passer rating; hence, it is useful to consider its associated margin of error. Given the complexity of the formula for the passer rating in terms of a quarterback's raw statistics, a simple expression (or any expression) for the margin of error is not available. Note that, because the season passer rating is not simply the average of the individual game passer ratings, it is not correct to use the formula for the margin of error of an average. Therefore, we consider a simulation-based approach.

The first step is to determine the level at which the simulation should take place. That is, should we simulate Flacco's 2012 season by simulating individual passing attempts, games, or something else? The basic idea in making this choice is to have the largest number of independent components on which to do the sampling. Because Flacco had 531 pass attempts but only 16 games, it might seem like passing attempts would be the better choice. However, the outcomes on passing attempts within a given game may be closely related. For instance, often the reason a passing play results in a touchdown is that the team had several successful passing plays on that drive; also, it is generally impossible to have two long passing plays on one drive. If a team is successful passing in the first half of the game, so that the team has a big lead, it might run relatively few, and more conservative, passing plays in the second half; on the other hand, if a quarterback throws an interception or two early in the game, he may need to throw more passes late in the game, making an additional interception more likely. Therefore, treating each passing play as an independent experiment might overstate the amount of variability in the passer rating. A better choice is to use games as the sampling unit.

To simulate Flacco's 2012 season, I took the results from the 16 regular season games in which he played (Dataset 4.2), took a random sample of 16 games, and calculated the season passer rating based on those results. Using a simulation sample size of 1000 (i.e., simulating the 2012 season 1000 times), I calculated the margin of error for Flacco's passer rating using the same general approach used for Durant's average PPG. The result is a margin of error of 12.9; that is, we can report Flacco's passer rating as 87.7 ± 12.9.

Table 4.6 contains the passer ratings and the corresponding margins of error for the 12 NFL quarterbacks with the highest passer ratings in 2012. Because of the weighting involved in constructing the passer rating, the margins of error tend to be fairly high, relative to the ratings themselves. The magnitude of the margin of error varies considerably with the quarterback because of the relative importance of factors contributing to a given quarterback's rating (yard per carry, completion percentage, touchdown percentage, and interception percentage) along with the game-to-game consistency of the quarterback. It is interesting to note that the quarterbacks with the smallest margin of error, Rodgers, Manning, Brady, and Ben Roethlisberger, are some of the most successful quarterbacks in the NFL, with 11 Super Bowl starts

**TABLE 4.6** Passer ratings and margins of error for top 2012 NFL quarterbacks

| PLAYER | PASSER RATING | MARGIN OF ERROR |
|---|---|---|
| Rodgers | 108.0 | 10.5 |
| Manning, P. | 105.8 | 10.7 |
| Griffin, III | 102.4 | 15.3 |
| Wilson | 100.0 | 16.0 |
| Ryan | 99.1 | 13.8 |
| Brady | 98.7 | 11.5 |
| Roethlisberger | 97.0 | 10.6 |
| Brees | 96.3 | 13.9 |
| Schaub | 90.7 | 12.0 |
| Romo | 90.5 | 13.3 |
| Rivers | 88.6 | 13.1 |
| Flacco | 87.7 | 12.9 |

between them; the other 8 quarterbacks in the list account for only 2 Super Bowl starts. The two rookie quarterbacks, Robert Griffin and Russell Wilson, have the highest margins of error.

# 4.6 COMPARISON OF TEAMS AND PLAYERS

Analytic methods are often concerned with comparing and contrasting players or teams. In making such comparisons, it is important to take into account the inherent variability in sports data. For example, suppose we are comparing the scoring of Durant and LeBron James in the 2011–2012 NBA regular season (Datasets 4.1 and 4.3, respectfully). According to game results from that season, Durant averaged 28.0 PPG, while James averaged 27.1 PPG. Based on these results, Durant clearly had a per game scoring average 0.9 PPG higher than that of James. However, can we conclude that Durant is a better scorer than James? Or, could the fact that Durant had a higher scoring average than James be because of the random nature of scoring in the NBA, and if we observed a very long sequence of games for both players, played under the conditions of the 2011–2012 season, would James actually have the higher scoring average?

To address this issue, we can take into consideration the margin of error of the 0.9-PPG difference in Durant's and James's scoring averages. Using the same approach used for Durant, the margin of error for James's scoring average is also 1.7. To find the margin of error for the difference between Durant's and James's scoring average, we use the following rule: The margin of error for the difference between two independent measurements is found by adding the squares of the two margins of error and taking the square root of the result. It follows that for the difference between Durant's and James's scoring average, the margin of error is

$$\sqrt{(1.7)^2 + (1.7)^2} = 2.4$$

and the difference between Durant's and James's scoring average can be reported as $0.9 \pm 2.4$ PPG. That is, our best guess of the true difference is 0.9 PPG, and we are reasonably certain that the true difference lies between −1.5 and 3.3 PPG. Note that this range includes negative values, suggesting that the true difference in scoring might favor James. Table 4.7 contains a similar comparison for the statistics presented in Table 4.1.

When comparing Durant's and James's statistics, we have chosen to base the comparisons on differences of their per game averages. An alternative approach is to consider the ratio of one player's average to the other player's average. For instance, to compare per game scoring, we could compute the ratio of Durant's PPG to James's PPG: 28.0/27.1 = 1.033, so that Durant's PPG is 3.3% higher than James's. Note that such an approach is generally only used for nonnegative statistics in which the average values are positive.

**TABLE 4.7**  Comparison of Durant and James based on 2011–2012 per game averages

|          | DURANT | JAMES | DIFFERENCE | MARGIN OF ERROR |
|----------|--------|-------|------------|-----------------|
| Rebounds | 8.0    | 7.9   | 0.0        | 1.0             |
| Assists  | 3.5    | 6.2   | -2.7       | 0.8             |
| Turnovers| 3.8    | 3.4   | 0.3        | 0.6             |
| Fouls    | 2.0    | 1.5   | 0.5        | 0.5             |
| Points   | 28.0   | 27.1  | 0.9        | 2.4             |

As with differences, it is important to take into account the margin of error of the ratio. Unfortunately, the formula for the margin of error of a ratio is a little more complicated than the formula for the margin of error of a difference. Consider two measurements $X$ and $Y$; let $\overline{X}$ and $\overline{Y}$ denote the corresponding sample means and let $R = \overline{X}/\overline{Y}$ denote the ratio. Then, the margin of error of $R$ is

$$R\sqrt{(ME(\overline{X})/\overline{X})^2 + (ME(\overline{Y})/\overline{Y})^2}$$

where $ME(\overline{X})$ and $ME(\overline{Y})$ denote the margins of error of $\overline{X}$ and $\overline{Y}$, respectively. Therefore, in the example, the margin of error is

$$(1.033)\sqrt{(1.7/28.0)^2 + (1.7/27.1)^2} = 0.090.$$

That is, the ratio of Durant's PPG to James's PPG is 1.033 with a margin of error of 0.090 so that the hypothetical true ratio is in the range 0.943 to 1.123; therefore, as with the difference-based analysis, the scoring averages of these two players are essentially the same.

When comparing parameter values, the comparison can be based either on differences or ratios, depending on which approach seems more natural for the quantity analyzed. Because scoring averages are more commonly compared by noting how many more PPG one player scores than another, in the Durant-James scoring example, a comparison based on differences seems more appropriate.

It is worth noting that when comparing parameters through their difference, it does not matter how the difference is calculated. For instance, Durant's PPG minus James's PPG is 0.9 with a margin of error of 1.4. If we look at James's PPG minus Durant's PPG, the result is −0.9 with a margin of error of 1.4, so we reach the same conclusion in either case. However, with ratios, the situation is a little different. The ratio of James's PPG to Durant's PPG is 0.968 with a margin of error of 0.084, so the true ratio is in the range 0.884 to 1.052. If we invert these values to obtain a range for the Durant-to-James ratio, the result is 0.951 to 1.131, which is close to, but not exactly the same as, the result obtained when we analyzed the Durant-to-James ratio directly. Fortunately, in many cases, the differences between the two analyses are small and, as in this case; do not affect the general conclusions.

# 4.7 COULD THIS RESULT BE ATTRIBUTED TO CHANCE? UNDERSTANDING STATISTICAL SIGNIFICANCE

Often, our goal in analyzing data is to draw conclusions about the players or teams generating the data. As part of this process, it is important to recognize the inherent randomness of the available data and that some features of the observed data might be consequences of this randomness. The term *statistically significant* is used to describe those conclusions that are unlikely to be attributed to chance.

For instance, consider the Durant-James scoring example. Durant's PPG average is greater than James's by 0.9 PPG. However, is the difference statistically significant? That is, suppose that their true scoring averages are actually the same. Is it likely that, based on the results of a single season (in fact, a shortened season), one player's PPG average will exceed the other player's PPG average by 0.9 or more points? This issue is addressed by considering the margin of error. Because the margin of error of the difference is 1.4 PPG, which is greater than the 0.9 PPG difference, we conclude that the difference between Durant's and James's scoring average is not statistically significant; that is, it may be a result of chance, and if a long sequence of games was observed, James might score more per game than Durant. For the five statistics for Durant and James compared in Table 4.7, only the difference in assists per game is statistically significant.

Now, consider a comparison of Blake Griffin and Durant. Griffin scored 20.7 PPG with a margin of error of 1.4 PPG. Therefore, the difference between Durant's and Griffin's PPG average is 7.3 PPG, with a margin of error of 2.2 PPG. Here, the margin of error is considerably less than the actual difference so that the range for the true difference, 5.1 to 9.5 PPG, consists entirely of positive values. Therefore, it is unlikely that the result that Durant had a higher scoring average than Griffin in 2011–2012 can be attributed to the inherent randomness in game-to-game scoring; that is, the difference is statistically significant.

When comparing two groups, statistical significance is assessed by asking whether the observed difference between groups is large relative to what might be expected because of "random variation." For example, consider the theory that there is a difference between the walks per 9 innings (BB9) of left-handed and right-handed pitchers. To investigate this issue, we use data from the 2009 MLB season, limiting the analysis to those pitchers with at least 30 innings pitched. The average value of BB9 for the 111 left-handed pitchers is 3.69; the average value for the 323 right-handed pitchers is 3.55. Therefore, there is a difference in the mean values for the two groups. However, if we just randomly partitioned the 434 pitchers into two groups, with 111 and 323 players, respectively, we know that the average values of BB9 for the two groups would not be exactly the same. The observed difference is considered statistically significant if it is large relative to what might be expected from such a random assignment to two groups.

To determine this, we compare the observed difference to the margin of error. The margin of error for the sample mean of BB9 for left-handed pitchers is 0.218; the corresponding value for right-handed pitchers is 0.126. It follows that the margin of error for the difference is about 0.25. Because the observed difference in the sample means (0.14) is less than the margin of error, we cannot rule out that it is a result of the inherent randomness of baseball; that is, the difference is not statistically significant, and based on this analysis, there is no evidence that there is a difference in the walk rates of left-handed and right-handed pitchers.

# 4.8 COMPARING THE AMERICAN AND NATIONAL LEAGUES

To illustrate the use and properties of statistical significance, in this section we consider several comparisons of the American League (AL) and National League (NL) in MLB. It is well known that the AL has dominated interleague play since it began in 1997. Of the 4264 regular season interleague games played from 1997 to 2013, the AL has won 2235, a proportion of 0.524. The margin of error of this proportion is

$$2\,\frac{\sqrt{0.524\,(1-0.524)}}{\sqrt{4264}}=0.015.$$

Because the interval $0.524 \pm 0.015 = (0.509, 0.539)$ does not include 0.500, we conclude that the observed winning proportion of the AL, 0.524, is statistically significantly different from 0.500. That is, such a large winning proportion is unlikely to have occurred simply by chance or luck.

In 2013, the AL won 154 of 300 interleague games, for a winning proportion of 0.513. This proportion has a margin of error of

$$2\,\frac{\sqrt{0.513\,(1-0.513)}}{\sqrt{300}}=0.058.$$

Because the interval $0.513 \pm 0.058 = (0.455, 0.571)$ includes 0.500, we cannot conclude that the observed winning proportion of the AL, 0.513, is statistically significantly different from 0.500. That is, if a very large number of interleague games had been played in 2013, based on the evidence available (the 300 games actually played), we cannot rule out the possibility that the NL would actually come out on top.

Note that the margin of error for the 2013 proportion is much larger than the margin of error for the proportion for the period 1997 to 2013. This is because the 2013 sample size (300) is much smaller than the sample size for 1997 to 2013 (4264). However, even though the sample size for 1997 to 2013 is about 14 times as large as the sample size for 2013, the margin of error in 2013 is only about four times as large as the margin of error for 1997 to 2013. This is because it is the square root of the sample size, not the sample size itself, that plays a role in the margin-of-error calculation. Note that the square root of 14 is about 3.8.

Other statistics for the two leagues can be compared using the same basic approach. For instance, consider home runs hit per game. AL teams played 2432 games in 2013 and averaged 1.030 HRs per game with a standard deviation of 1.081; the data in this section are from Baseball-Reference.com. Therefore, the margin of error of AL average home runs per game is

$$2\frac{1.081}{\sqrt{2432}} = 0.044.$$

National League teams played 2430 games; for these teams, the average HRs per game is 0.888, and the standard deviation is 1.006. The margin of error of the NL average HRs per game is

$$2\frac{1.006}{\sqrt{2430}} = 0.041.$$

The difference between the average HRs per game for the AL and the NL is

$$1.030 - 0.888 = 0.142$$

with a margin of error of

$$\sqrt{(0.044)^2 + (0.041)^2} = 0.060.$$

Note that because the two margins of error are approximately the same, the margin of error of the difference is approximately $\sqrt{2} = 1.4$ times either one. Since the difference in the average HRs per game (0.142) is greater than the margin of error, we can conclude that the difference is statistically significant. That is, AL teams tend to hit more HRs per game than do NL teams.

Because NL batters face NL pitchers much more often than they face AL pitchers, this result could be because AL pitchers are weaker than NL pitchers, at least with respect to home runs allowed. Comparing the home runs allowed of AL and NL pitchers is potentially misleading for the same reason: AL pitchers primarily face AL batters, and NL pitchers primarily face NL batters. Therefore, to compare AL and NL pitching, we should compare the home runs allowed of AL and NL teams against AL and NL batters separately.

For instance, consider games versus AL batters during the 2013 MLB season. AL teams allowed an average of 1.042 home runs per game in 2132 games, with a standard deviation of 1.084 HRs. It follows that the margin of error of this average is

$$2\frac{1.084}{\sqrt{2132}} = 0.047.$$

NL teams allowed an average of 0.943 home runs per game in games versus AL batters, that is, in interleague games, with a standard deviation of 1.052 home runs. Because there were 300 interleague games, the margin of error of this average is

$$2\,\frac{1.052}{\sqrt{300}} = 0.121.$$

The difference in HRs allowed per game versus AL batters for AL and NL teams is 0.099 HRs per game, with a margin of error of

$$\sqrt{(0.047)^2 + (0.121)^2} = 0.136.$$

Therefore, AL teams gave up more home runs per game than NL teams in games versus AL batters. However, because the difference in average home runs allowed per game (0.099) is smaller than its margin of error (0.136), we conclude that the difference is not statistically significant, and there is no statistical evidence of a difference in HRs allowed by AL teams and NL teams in games versus AL batters.

In the same way, we can compare the home runs allowed by AL and NL teams versus NL batters; the calculations are analogous to the ones comparing NL and AL teams versus AL batters, so only the results are given. NL teams gave up an average of 0.877 home runs per game versus NL batters; the margin of error of this average is 0.043. AL teams allowed an average of 0.960 home runs per game versus NL batters, with a margin of error of 0.128. Therefore, AL teams allowed 0.083 more home runs per game versus NL batters than did NL teams; however, the margin of error of this difference is 0.135, so that it is not statistically significant. These results suggest that AL pitchers are, in fact, weaker than NL pitchers in terms of home runs allowed; however, the results are not conclusive because the differences in home runs allowed are not statistically significant. Also, these results do not adjust for other factors, such as park effects, which might be important.

Note that the margins of error in these comparisons are much larger than the margin of error when comparing the home runs per game of AL and NL teams. This is because, in each case, one of the sample sizes, the one based on interleague games, is relatively small (300). Although the other sample size in each comparison is large (2130 or 2132), one small sample size is enough to inflate the margin of error for the comparison. That is, the margin of error of the difference of two statistics is small only if the margins of error of both statistics are small.

Finally, we might compare the home runs per game of AL and NL teams in interleague games. This would eliminate park effects, as well as some other factors that may have affected the comparison. In interleague games, AL teams hit an average of 0.943 home runs per game, with a margin of error of 0.121; in these games, NL teams hit an average of 0.960 home runs per game, with a margin of error of 0.128. Therefore, the NL teams had a slight advantage of 0.017 home runs per game; however, with the large margin of error, about 0.18, the difference is clearly not statistically significant.

# 4.9 MARGIN OF ERROR AND ADJUSTED STATISTICS

Adjusted statistics, which allow us to account for the effects of playing conditions, opponents, coaches' decisions, and so on, play an important role in the analysis of sports data. However, it is important to keep in mind that all such adjustments are the results of statistical procedures, and although they often yield adjusted statistics that are improvements over the original ones, the process of adjustment generally introduces additional variability into the statistic, in the sense that the margin of error of the adjusted statistic is larger than that of the original, unadjusted statistic.

First consider "direct adjustment," as discussed in Sections 3.11 and 3.12. This method applies to statistics that can be viewed as weighted averages over specific subclasses and proceeds by reweighting these subclass-specific statistics according to some standard weights. However, this method has the potential drawback of giving large weight to a statistic based on few data, leading to a large margin of error for the adjusted statistic.

Consider the case of Jason Hanson, kicker for the Detroit Lions in 2011. That year, he had 9 attempts in the 20- to 29-yard range, making all 9; 9 attempts in the 30- to 39-yard range, making 8; 4 attempts in the 40- to 49-yard range, making 2; and 7 attempts of 50 yards or greater, making 5. Overall, he made 24 field goals in 29 attempts, for a success proportion of 0.828. Using a simple binomial model for his field goal attempts, the margin of error of his success proportion is

$$2\,\frac{\sqrt{0.828\,(1-0.828)}}{29} = 0.140;$$

using a more sophisticated approach that takes into account the fact that his success proportion is different in different ranges, the margin of error is 0.133.

Note that Hanson's overall field goal success ratio can be written in terms of the distance-specific success ratios, together with the weights based on the number of attempts, as follows:

$$\frac{24}{29} = \frac{9}{29}\frac{9}{9} + \frac{9}{29}\frac{8}{9} + \frac{4}{29}\frac{2}{4} + \frac{7}{29}\frac{5}{7}.$$

Therefore, the success ratio in the 40- to 49-yard range, which is based on only 4 attempts, naturally receives little weight.

Now, consider the adjusted field goal proportion, as presented in Section 3.12. For Hanson, this is

$$0.778 = (0.291)\frac{9}{9} + (0.265)\frac{8}{9} + (0.306)\frac{2}{4} + (0.138)\frac{5}{7}.$$

Note that the results from the 40- to 49-yard range, based on only 4 attempts, receives the largest weight in this computation. This leads to a relatively large margin of error of 0.175 for the adjusted statistic; this is about 1.25 times the margin of error for the original statistic. Therefore, although the adjustment for the distance of a kicker's attempts makes the success proportions more directly comparable, it has the additional effect of introducing additional error into the comparisons.

Similar considerations apply to other types of adjustment. For instance, let $Y$ denote some statistics, and suppose we observe $Y_0$, the value of that statistic in a given year ("year 0"). Suppose that we want to adjust this statistic to obtain an "equivalent" value for a different year ("year 1"). Let $Y_1$ denote the (unobserved) statistic for that player in year 1. To estimate $Y_1$, we need to make some assumption about how $Y_0$ and $Y_1$ are related. For instance, we might assume that

$$\frac{Y_0}{\overline{Y}_0} = \frac{Y_1}{\overline{Y}_1}$$

where $\overline{Y}_0, \overline{Y}_1$ denote the league averages of the statistic for years 0 and 1, respectively. Under this assumption, the ratio of a player's statistic to the league average is constant over years. It follows that the adjusted value of $Y_0$, adjusted to year 1, is

$$\frac{\overline{Y}_1}{\overline{Y}_0} Y_0.$$

For instance, two of the highest values of OPS (on-base plus slugging) for a season are Bobby Bonds's 1.422 in 2004 and Babe Ruth's 1.379 in 1920. Direct comparison of these values is distorted by the differences in baseball in 1920 and 2004; hence, we might consider adjusting these values to some standard year, which we take to be 2012. In 2012, the MLB average for OPS is 0.724; in 2004, it is 0.763, and in 1920 it is 0.707. Therefore, Bonds's 2004 OPS, adjusted to 2012, is

$$\frac{0.724}{0.763} 1.422 = 1.349;$$

Ruth's 1920 OPS, adjusted to 2012, is

$$\frac{0.724}{0.707} 1.379 = 1.412.$$

It follows that Ruth's 1920 season is more impressive than Bonds's 2004 season, at least in terms of OPS and the type of adjustment used here.

In this type of multiplicative adjustment, the adjusted statistic is simply the original statistic multiplied by a factor $f$. For concreteness, assume that $f$ has the form

$$f = \frac{\overline{Y}_1}{\overline{Y}_0},$$

where $\overline{Y}_0$, and $\overline{Y}_1$ are the league averages of the statistic for years 0 and 1, respectively, as in the Ruth and Bonds example.

Because the adjusted statistic is simply the original statistic multiplied by $f$, the margin of error of the adjusted statistic is $f$ times the margin of error of the original statistic plus an additional factor because the adjustment factor $f$ is a function of data; hence, it has its own margin of error. This additional factor is small if the margins of error of $\overline{Y}_0$, $\overline{Y}_1$ are small relative to the margin of error of $Y_0$. For instance, if the number of observations on which each of $\overline{Y}_0$, $\overline{Y}_1$ is based is at least 20 times as great as the number of observations on which $Y_0$ is based, then the contribution of this additional factor is generally less than 5%, which might be considered to be negligible. When $Y_0$ is a statistic for a given player or team and $\overline{Y}_0$, $\overline{Y}_1$ are the corresponding statistics for the league, this condition is usually satisfied.

The lesson here is that, when making adjustments to statistics, it is important for the adjustment factors to be accurately determined. To achieve this, it is tempting to use as many data as possible, perhaps including results for several years or leagues. Although this approach is often valid, it is important to keep in mind that, unless the relationships between the variables are constant for the data being used, it has the potential of introducing additional error. Therefore, the analyst must balance the benefits of using more data to obtain more accuracy versus possible biases introduced by expanding the range of the analysis.

# 4.10 IMPORTANT CONSIDERATIONS WHEN APPLYING STATISTICAL METHODS TO SPORTS

Statistical methods are based on a formal model for the process generating the data being analyzed. For instance, let $X_1,\ldots,X_n$ denote the available data. The methods discussed in this chapter assume that $X_1,\ldots,X_n$ are the observed values of independent random variables each with the same distribution. Let $X$ denote a random variable that has the same distribution as $X_1,\ldots,X_n$. For instance, consider the Durant scoring example in which $X$ represents Durant's points scored for a given game in the 2011–2012 season and $X_1,\ldots,X_n$ represent his actual points scored in the 66 games of the season (i.e., $n = 66$).

Note that this formulation involves a number of assumptions. For instance, we assume that each $X_j$ has the same distribution, and that, for any $i$ and $j$, $X_i$ and $X_j$ are independent. In the context of the Durant scoring example, this means that the random variable representing Durant's scoring for a given game has the same distribution for each game. This assumption is unlikely to be literally true because his scoring might depend on the opponent, the location of the game (home or away), and so on. Therefore, the concept of a "long-run scoring average" is a hypothetical one. Also, the assumption of independence might not hold if, for example, the results of one game influence how Durant plays in the following game.

In many applications of statistics, assumptions of this type are addressed through the use of random sampling to collect data. However, we know that with sports data random sampling is not an option. In some cases, additional data can be collected that allow us to evaluate, and possibly eliminate, assumptions. For instance, if we believe that Durant's scoring is much different for home and away games, we could analyze home and away games separately. Chapters 5–7 are devoted to modeling the relationships between variables and those methods can be used in this context. Even so, it is safe to say that many of the standard assumptions used in statistics will rarely be satisfied in our setting.

That being said, concepts such as the margin of error are still useful for understanding the role of randomness in sports statistics. However, such concepts should be viewed as guidelines rather than as strict results.

## 4.11 COMPUTATION

The majority of the calculations in this section are easily performed using a calculator or basic spreadsheet functions. An exception is the simulation method described in Section 4.5. Here, a basic approach to obtaining simulation results is outlined. Readers with more extensive programming experience will have no trouble automating the procedure further.

Consider using simulation to estimate the margin of error of Joe Flacco's passer rating for the 2012 NFL season. Figure 4.1 shows a spreadsheet of Flacco's game-by-

| | A | B | C | D | E | F |
|---|---|---|---|---|---|---|
| 1 | Game | Comp | Att | Yds | TD | Int |
| 2 | 1 | 21 | 29 | 299 | 2 | 0 |
| 3 | 2 | 22 | 42 | 232 | 1 | 1 |
| 4 | 3 | 28 | 39 | 382 | 3 | 1 |
| 5 | 4 | 28 | 46 | 356 | 1 | 1 |
| 6 | 5 | 13 | 27 | 187 | 0 | 1 |
| 7 | 6 | 17 | 26 | 234 | 1 | 0 |
| 8 | 7 | 21 | 43 | 147 | 1 | 2 |
| 9 | 8 | 15 | 24 | 153 | 1 | 0 |
| 10 | 9 | 21 | 33 | 341 | 3 | 1 |
| 11 | 10 | 20 | 32 | 164 | 0 | 0 |
| 12 | 11 | 30 | 51 | 355 | 1 | 0 |
| 13 | 12 | 16 | 34 | 188 | 1 | 1 |
| 14 | 13 | 16 | 21 | 182 | 3 | 1 |
| 15 | 14 | 20 | 40 | 254 | 2 | 1 |
| 16 | 15 | 25 | 36 | 309 | 2 | 0 |
| 17 | 16 | 4 | 8 | 34 | 0 | 0 |
| 18 | Total | 317 | 531 | 3817 | 22 | 10 |
| 19 | Rating | | 87.7 | | | |
| 20 | | | | | | |

**FIGURE 4.1**   A spreadsheet of Flacco's game-by-game results.

game results together with his season totals and his passer rating, calculated using the formula discussed in Section 4.5.

To construct a "simulated season" for Flacco, we take a random sample of his 2012 game results. Because these results are contained in rows 2 through 17 of the spreadsheet, this can be achieved by choosing a random integer from 2 to 17 and then recording the corresponding row of the spreadsheet. Performing this 16 times yields a simulated 16-game season. Using the results of this simulated season, we can calculate a simulated passer rating for Flacco.

To calculate the random row number in Excel, we can use the command

*RANDBETWEEN*(2, 17)

which returns a random integer in the range 2 to 17 (including possibly 2 and 17). To extract the corresponding data from the spreadsheet of Flacco's game-by-game results, we use the INDIRECT command in Excel, which allows for indirect cell references using text strings. For instance, suppose that the random integer specifying the desired row of the data table is in cell H2. The command

*INDIRECT*("*B*"&*$H*2)

returns the value of the cell in column B and row given by H2. By using this command with columns B, C, D, E, and F we can extract the data corresponding to the row given by the value in H2; when H2 contains a random number between 2 and 17, then this procedure yields the data from the game given by the random number.

Figure 4.2 contains a simulated season using this approach. Column H contains the row numbers corresponding to the simulated season, obtained using the RANDBETWEEN function. For instance, the first row number obtained is 6, corresponding to game 5 in Flacco's actual 2012 season. Columns J though N of row 2 contain the data from columns B through F of row 6, extracted using the INDIRECT function as discussed previously. This procedure is performed 16 times to obtain a simulated season for Flacco. The game-by-game statistics are totaled to obtain a simulated season, and based on these results, a simulated passer rating for Flacco is calculated using the usual formula; for the random numbers obtained in this example, this simulated passer rating is 77.0.

To estimate the margin of error of Flacco's passer rating, we need to repeat this procedure several times. Using the recalculation feature of Excel, new random numbers are chosen, leading to new rows being extracted from the original data, which results in a new simulated passer rating. This type of recalculation can be triggered by pressing F9; it is also triggered by other operations on the spreadsheet, which can be useful in further automating the procedure. Figure 4.3 contains a new simulated season obtained after recalculation.

Flacco's performance in the second simulated season is much better than in the first simulated season (and better than his actual season), with 4300 yards, 26 touchdowns, and a passer rating of 98.2. The simulated values obtained this way give us some indication of the natural variability in Flacco's passer rating. Repeating the simulation several times and calculating the standard deviation of the resulting simulated passer ratings can be used to estimate the margin of error of Flacco's passer rating, as discussed in Section 4.5.

| | A | B | C | D | E | F | G | H | I | J | K | L | M | N |
|---|---|---|---|---|---|---|---|---|---|---|---|---|---|---|
| 1 | Game | Comp | Att | Yds | TD | Int | | Row | Game | Comp | Att | Yds | TD | Int |
| 2 | 1 | 21 | 29 | 299 | 2 | 0 | | 6 | 5 | 13 | 27 | 187 | 0 | 1 |
| 3 | 2 | 22 | 42 | 232 | 1 | 1 | | 5 | 4 | 28 | 46 | 356 | 1 | 1 |
| 4 | 3 | 28 | 39 | 382 | 3 | 1 | | 13 | 12 | 16 | 34 | 188 | 1 | 1 |
| 5 | 4 | 28 | 46 | 356 | 1 | 1 | | 12 | 11 | 30 | 51 | 355 | 1 | 0 |
| 6 | 5 | 13 | 27 | 187 | 0 | 1 | | 2 | 1 | 21 | 29 | 299 | 2 | 0 |
| 7 | 6 | 17 | 26 | 234 | 1 | 0 | | 8 | 7 | 21 | 43 | 147 | 1 | 2 |
| 8 | 7 | 21 | 43 | 147 | 1 | 2 | | 3 | 2 | 22 | 42 | 232 | 1 | 1 |
| 9 | 8 | 15 | 24 | 153 | 1 | 0 | | 16 | 15 | 25 | 36 | 309 | 2 | 0 |
| 10 | 9 | 21 | 33 | 341 | 3 | 1 | | 17 | 16 | 4 | 8 | 34 | 0 | 0 |
| 11 | 10 | 20 | 32 | 164 | 0 | 0 | | 13 | 12 | 16 | 34 | 188 | 1 | 1 |
| 12 | 11 | 30 | 51 | 355 | 1 | 0 | | 12 | 11 | 30 | 51 | 355 | 1 | 0 |
| 13 | 12 | 16 | 34 | 188 | 1 | 1 | | 3 | 2 | 22 | 42 | 232 | 1 | 1 |
| 14 | 13 | 16 | 21 | 182 | 3 | 1 | | 8 | 7 | 21 | 43 | 147 | 1 | 2 |
| 15 | 14 | 20 | 40 | 254 | 2 | 1 | | 7 | 6 | 17 | 26 | 234 | 1 | 0 |
| 16 | 15 | 25 | 36 | 309 | 2 | 0 | | 15 | 14 | 20 | 40 | 254 | 2 | 1 |
| 17 | 16 | 4 | 8 | 34 | 0 | 0 | | 9 | 8 | 15 | 24 | 153 | 1 | 0 |
| 18 | Total | 317 | 531 | 3817 | 22 | 10 | | Total | | 321 | 576 | 3670 | 17 | 11 |
| 19 | Rating | | 87.7 | | | | | Rating | | 77.0 | | | | |
| 20 | | | | | | | | | | | | | | |

**FIGURE 4.2**   A simulated season for Flacco.

| | A | B | C | D | E | F | G | H | I | J | K | L | M | N |
|---|---|---|---|---|---|---|---|---|---|---|---|---|---|---|
| 1 | Game | Comp | Att | Yds | TD | Int | | Row | Game | Comp | Att | Yds | TD | Int |
| 2 | 1 | 21 | 29 | 299 | 2 | 0 | | 6 | 5 | 13 | 27 | 187 | 0 | 1 |
| 3 | 2 | 22 | 42 | 232 | 1 | 1 | | 4 | 3 | 28 | 39 | 382 | 3 | 1 |
| 4 | 3 | 28 | 39 | 382 | 3 | 1 | | 10 | 9 | 21 | 33 | 341 | 3 | 1 |
| 5 | 4 | 28 | 46 | 356 | 1 | 1 | | 5 | 4 | 28 | 46 | 356 | 1 | 1 |
| 6 | 5 | 13 | 27 | 187 | 0 | 1 | | 5 | 4 | 28 | 46 | 356 | 1 | 1 |
| 7 | 6 | 17 | 26 | 234 | 1 | 0 | | 4 | 3 | 28 | 39 | 382 | 3 | 1 |
| 8 | 7 | 21 | 43 | 147 | 1 | 2 | | 2 | 1 | 21 | 29 | 299 | 2 | 0 |
| 9 | 8 | 15 | 24 | 153 | 1 | 0 | | 11 | 10 | 20 | 32 | 164 | 0 | 0 |
| 10 | 9 | 21 | 33 | 341 | 3 | 1 | | 17 | 16 | 4 | 8 | 34 | 0 | 0 |
| 11 | 10 | 20 | 32 | 164 | 0 | 0 | | 15 | 14 | 20 | 40 | 254 | 2 | 1 |
| 12 | 11 | 30 | 51 | 355 | 1 | 0 | | 16 | 15 | 25 | 36 | 309 | 2 | 0 |
| 13 | 12 | 16 | 34 | 188 | 1 | 1 | | 15 | 14 | 20 | 40 | 254 | 2 | 1 |
| 14 | 13 | 16 | 21 | 182 | 3 | 1 | | 10 | 9 | 21 | 33 | 341 | 3 | 1 |
| 15 | 14 | 20 | 40 | 254 | 2 | 1 | | 2 | 1 | 21 | 29 | 299 | 2 | 0 |
| 16 | 15 | 25 | 36 | 309 | 2 | 0 | | 17 | 16 | 4 | 8 | 34 | 0 | 0 |
| 17 | 16 | 4 | 8 | 34 | 0 | 0 | | 16 | 15 | 25 | 36 | 309 | 2 | 0 |
| 18 | Total | 317 | 531 | 3817 | 22 | 10 | | Total | | 327 | 521 | 4301 | 26 | 9 |
| 19 | Rating | | 87.7 | | | | | Rating | | 98.2 | | | | |

**FIGURE 4.3**   A second simulated season for Flacco.

# 4.12 SUGGESTIONS FOR FURTHER READING

Like the topics in Chapter 2, the topics covered in this chapter are covered in many books on statistics; see, for example, the works of Agresti and Finlay (2009), McClave and Sinich (2006), and Moore and McCabe (2005). Mlodinow (2008) gives non-technical descriptions of margins of error (Chapter 7) and statistical significance (Chapter 9). The appendix of the work of Tango, Lichtman, and Dolphin (2007) discusses calculation and approximation of margins of error for some estimates commonly used in analyzing sports data.

The simulation-based method presented in Section 4.5 is known as the bootstrap. The bootstrap is a powerful method that substitutes computing power for analytic calculations; Freedman (2009, Chapter 8) provides a useful introduction to this important area of statistics.

A detailed comparison of Bonds and Ruth is given by Silver (2006a).

# Using Correlation to Detect Statistical Relationships

# 5

## 5.1 INTRODUCTION

One of the primary goals of analytic methods is to understand the relationships between variables. For instance, we might be interested in the relationship between player performance and team performance or in the relationship between performance in a given year and performance in the following year. In this chapter, several different approaches for measuring the strength of the relationship between variables are presented; these measures reduce the properties of such a relationship to a single number that is useful as a simple summary of the relationship between variables. In Chapter 6, we go a step further and consider mathematical models for these relationships.

## 5.2 LINEAR RELATIONSHIPS: THE CORRELATION COEFFICIENT

Suppose that for each subject, such as a player or team, we measure two variables $X$ and $Y$ and suppose that we are interested in the relationship between these variables. For instance, if the subjects are MLB (Major League Baseball) players, for each player we might record his number of hits ($X$) and his number of runs scored ($Y$) in a given season. The simplest way to gain some insight into the relationship between $X$ and $Y$ is to construct a scatterplot, in which the pair ($X$, $Y$) is plotted for each subject. Figure 5.1 contains a plot of runs scored versus hits for 2011 MLB players with a qualifying number of plate appearances (502 or more); there are 145 such players, so the sample size is $n = 145$. This dataset was analyzed previously in Chapter 2 (Dataset 2.4).

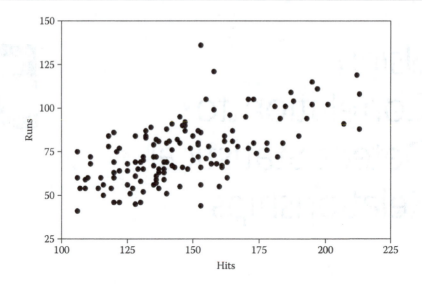

**FIGURE 5.1**    Runs versus hits for 2011 MLB players.

Like the plot in Figure 5.1, scatterplots often show a general linear relationship between the variables. In some cases, the pattern is vague, with considerable variation around the underlying linear trend; in other cases, the data values might come close to falling on a specific line. That is, in some cases, the relationship is strong so that the value of $X$ almost completely determines the value of $Y$; in other cases, the relationship is weak, with the value of $X$ giving, at best, a general indication regarding the value of $Y$. For instance, the relationship between runs and hits, as given in Figure 5.1, is a fairly strong linear relationship, and the relationship between walks and hits for qualifying 2011 MLB players (also in Dataset 2.4), given in the plot in Figure 5.2, shows only a weak linear relationship between the variables.

The correlation coefficient, denoted by $r$, is a single-number measure of the extent to which data cluster around a line. For instance, for runs and hits, as given in Figure 5.1, $r = 0.64$; for walks and hits, as given in Figure 5.2, $r = 0.17$. Figures 5.3–5.6 give several examples of scatterplots. Figure 5.3 contains a plot of offensive rebounds per game versus defensive rebounds per game for NBA (National Basketball Association) players with at least 56 games played in the 2011–2012 season (Dataset 5.1); here, $r = 0.69$ ($n = 182$). Figure 5.4 contains a plot of touchdown passes per passing attempt versus sacks per passing attempt for 2009 NFL (National Football League) quarterbacks with at least 160 passing attempts (Dataset 5.2); here, $r = -0.41$ ($n = 30$). Figure 5.5 contains a plot of 2011 wins versus 2010 wins for MLB teams (Dataset 5.3); here, $r = 0.41$ ($n = 30$). Figure 5.6 contains a plot of points scored versus points allowed for 2011 NFL teams (Dataset 5.4); here, $r = -0.07$ ($n = 32$). Note that the larger the value of $|r|$, the more closely the variables follow a linear relationship, with the sign of $r$ indicating the direction of the relationship. Computation of the correlation coefficient is discussed in Section 5.15.

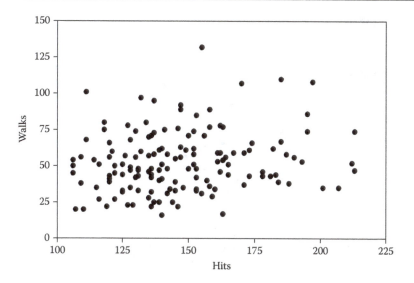

**FIGURE 5.2** Walks versus hits for 2011 MLB players.

The key to effectively using the correlation coefficient is to understand its properties. Let $r$ denote a correlation coefficient $r$. Then,

- $-1 \le r \le 1$; that is, $r$ always lies between $-1$ and $1$.
- $r > 0$ indicates an increasing linear relationship in which larger values of $X$ are associated with larger values of $Y$.

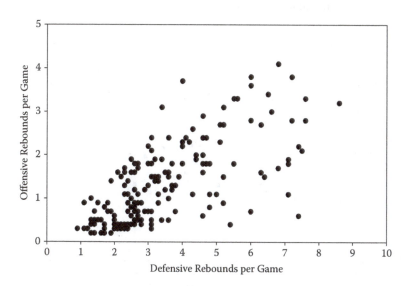

**FIGURE 5.3** Offensive versus defensive rebounds for 2011–2012 NBA players.

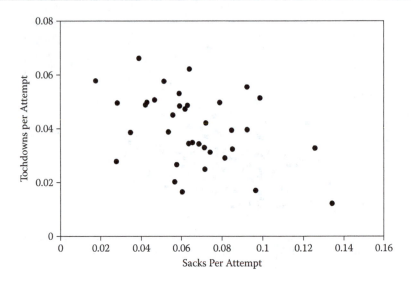

**FIGURE 5.4**   Touchdowns versus sacks for 2009 NFL quarterbacks.

- $r < 0$ indicates a decreasing linear relationship in which larger values of $X$ are associated with smaller values of $Y$.
- $r = 0$ indicates that there is no linear relationship between the variables.
- $r = 1$ or $r = -1$ indicates a perfect linear relationship between the variables.
- $r$ is not affected by switching the roles of the variables; for example, in Figure 5.1, if "runs" becomes the $X$-variable and "hits" becomes the $Y$-variable, the value of $r$ is unchanged.

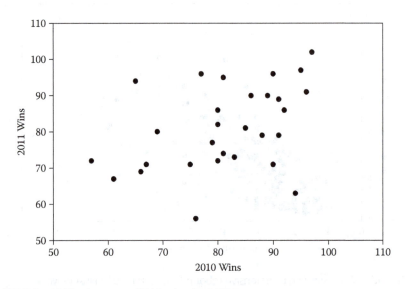

**FIGURE 5.5**   2011 wins versus 2010 wins for MLB teams.

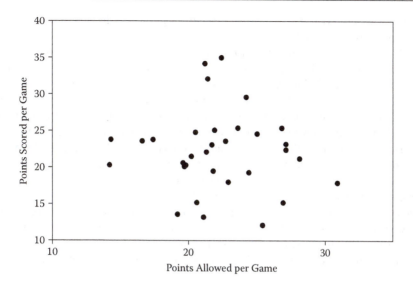

**FIGURE 5.6**   Points scored versus points allowed for 2011 NFL teams.

- Adding a constant to a variable does not change the value of $r$. For instance, in Figure 5.5, if instead of comparing 2011 wins and 2010 wins of MLB teams, we compare wins minus 81, so that 90 wins is recorded as 9 and 60 wins is recorded as –21, for example, the correlation is unchanged.
- $r$ is not affected by changing the units of the variables; for example, in Figure 5.6, if we analyze total points per season rather than points per game, the value of $r$ is unchanged. Note that this holds because each team played exactly 16 games. In Figure 5.3, if we analyze total rebounds instead of rebounds per game, the value of $r$ changes slightly (to 0.70) because there is some variation in the number of games each player plays.

Therefore, values of $r$ near 0 indicate a weak linear relationship, if any at all, with larger values of $|r|$ indicating stronger linear relationships. In particular, if the random variables under consideration are independent, we expect $r$ to be close to 0; it is unlikely to be exactly 0 because of the random nature of the data. Assessments of the magnitude of $|r|$ generally depend on the nature of the variables being considered. In well-controlled scientific experiments, values of $|r|$ close to 1 might be observed; however, with sports data, values of $|r|$ greater than 0.9 are relatively rare.

Interest often centers on whether two variables have a linear relationship. Because a value of $r$ of exactly 0 is virtually impossible with real data, it is natural to ask if a nonzero correlation coefficient is statistically significantly different from 0. A simple way to assess statistical significance is to compare $|r|$ to $2/\sqrt{n}$, where $n$ denotes the number of subjects under consideration. If $|r| < 2/\sqrt{n}$, we conclude that, even though it is not exactly 0, there is not a statistically significant linear relationship between the variables. That is, the observed linear relationship may be attributed to chance. For the examples illustrated by Figures 5.1–5.6, the correlation between the points scored and points allowed of NFL teams is not statistically significant; the other correlations are all statistically significant.

When evaluating significance in this way, there are some important issues to keep in mind; although we discuss these in the context of correlation, they apply to statistical significance in general. Suppose that the sample size is very large; then, an observed correlation can be small but still statistically significant. For example, if $n = 5000$, a correlation as small as 0.03 is still statistically significant. This means that, if the true correlation is actually 0, it is unlikely that we would observe a correlation with magnitude as large as 0.03. However, such a small correlation is unlikely to be practically important. On the other hand, if $n$ is small even relatively large correlations may not be statistically significant. For instance, if $n = 20$, even a correlation as large as 0.44 is not statistically significant. In these cases, it may be that there is a true correlation that is large enough to be important, but the sample size is too small to be able to say that with certainty. If possible, collecting more data will allow us to reduce this uncertainty; otherwise, we can use the observed correlation in the analysis but use caution in drawing firm conclusions.

The margin of error of a correlation coefficient $r$, based on a sample of size $n$, is given by

$$2\frac{\sqrt{1-r^2}}{\sqrt{n}}.$$

For example, for the correlation between the number of hits and number of runs of MLB players, considered previously in this section, $r = 0.64$, with a margin of error of 0.13.

One of the appeals of the correlation coefficient is its simplicity—it reduces the possibly complex relationship between two variables to a single number. But, such simplicity has some drawbacks; hence, there are some important issues to keep in mind when basing conclusions on correlations.

One is that the correlation coefficient considers only one type of relationship, a linear one. Therefore, a correlation coefficient near 0 does not imply that the variables are not related, only that there is no evidence of a linear relationship. Although in some fields nonlinear relationships arise naturally and are common, they are relatively uncommon when analyzing sports data; methods of analyzing certain types of nonlinear relationships are considered in Sections 5.4 and 6.8. Therefore, although a small value of $|r|$ often indicates that the variables are not related, to be safe, one should always supplement the correlation coefficient with a scatterplot to rule out the possibility of an important nonlinear relationship. In the remainder of the book, although such plots are not always presented, they have been examined for the presence of a nonlinear relationship.

Another important fact is that even if $X$ and $Y$ are correlated, that is, they have a statistically significant correlation, that does not imply that $X$ and $Y$ have any type of cause-and-effect relationship. In particular, an observed linear relationship between $Y$ and $X$ might be because both $X$ and $Y$ are related to a third variable.

Consider NFL teams during the 2010 and 2011 seasons. Let $X$ denote the draft order of the team for the 2011 draft so that, for Carolina, $X = 1$ (they took Cam Newton), and for the Super Bowl champion Green Bay Packers, $X = 32$. Let $Y$ denote the number of wins in the 2011 season, the season following the 2011 draft; these data are available in Dataset 5.5. Then, the correlation of $X$ and $Y$ is 0.36, suggesting that having later draft picks corresponds to more wins in the following season. If the relationship is a causal one, then teams with early first-round draft picks should trade for picks later in the round.

Of course, such a causal relationship between draft order and wins in the following season is unlikely; a more likely explanation is that both variables are affected by a third variable, the number of wins in 2010. If a team has few wins in 2010, that team has a high draft pick, and we expect that the team will not win many games in 2011 (although they may win more than in 2010). Let $Z$ be the number of wins in 2010. Because both $X$ and $Y$ are related to $Z$, they are related to each other. This type of "lurking variable" is discussed further in Section 5.6.

Finally, it is important to keep in mind that the correlation coefficient is only a single number; therefore, often it does not tell the whole story regarding the relationship between the variables, even when the relationship is a linear one. For instance, consider Figure 5.3 on the relationship between offensive and defensive rebounds of NBA players. The correlation is 0.69, which accurately describes a reasonably strong linear relationship between the variables. However, another interesting aspect of the plot of the data is that there is much less variability in the relationship between the variables when both values are small than when both variables are large. That is, players with few defensive rebounds almost always have few offensive rebounds; on the other hand, players with many defensive rebounds not only often have many offensive rebounds but also often have relatively few offensive rebounds. This is an important fact regarding the relationship between offensive and defensive rebounds that is not addressed by simply looking at the correlation coefficient; that is, the correlation coefficient is no substitute for examining a plot of the data.

## 5.3 CAN THE "PYTHAGOREAN THEOREM" BE USED TO PREDICT A TEAM'S SECOND-HALF PERFORMANCE?

The "Pythagorean formula" in baseball, first proposed by Bill James, is an attempt to relate a team's runs scored and runs allowed to its win-loss record. Let $F$ denote a team's runs scored and let $A$ denote the team's runs allowed. James's original result stated a team's winning percentage can be approximated by its "Pythagorean winning percentage," given by

$$\frac{F^2}{F^2 + A^2};$$

the term *Pythagorean* in this context is because of the formula's similarity to the Pythagorean theorem of elementary geometry. The exponent 2 in this expression was further refined to 1.83, leading to the expression

$$\frac{F^{1.83}}{F^{1.83} + A^{1.83}}.$$

Estimation of the exponent is considered in Section 6.10. The book *Baseball between the Numbers* (Keri, 2006) contains interesting discussion on the background and use of the Pythagorean formula in baseball.

For instance, in 2012, the Detroit Tigers scored 726 runs and allowed 670; it follows that their Pythagorean winning percentage was

$$\frac{(726)^{1.83}}{(726)^{1.83} + (670)^{1.83}} = 0.537;$$

their actual winning percentage was 0.543. Teams whose winning percentage exceeds their Pythagorean winning percentage might be considered to be "lucky," winning more games than they "should" based on the number of runs they scored and allowed.

It is sometimes claimed that the Pythagorean winning percentage is a better predictor of the future performance of a team than is the actual winning percentage; see Winston's work (2009, Chapter 1) for some anecdotal evidence. To investigate this possibility, we can conduct the following study: For each team during the 2008–2012 MLB seasons, the Pythagorean winning percentage based on the first half of the season (i.e., games played before the All-Star break) was calculated. This value can then be compared to the team's second-half winning percentage; these data are available in Dataset 5.6.

For instance, in 2012, in the first half of the season, the Los Angeles Dodgers' Pythagorean winning percentage was 0.514, and their actual first-half winning percentage was 0.540; their second-half winning percentage was 0.520.

Carrying out these calculations for each team, the correlation between the first-half Pythagorean winning percentage and the second-half actual winning percentage for the 150 teams (30 teams in each of five seasons) is 0.458. For comparison, the correlation between the actual first-half winning percentage and the second-half winning percentage is slightly higher, 0.472. Therefore, based on these results, the linear relationship between first-half Pythagorean winning percentage and second-half winning percentage is no stronger than is the linear relationship between the first-half winning percentage and second-half winning percentage.

# 5.4  USING RANK CORRELATION FOR CERTAIN TYPES OF NONLINEAR RELATIONSHIPS

One drawback of the correlation coefficient is that it measures a specific type of relationship, a linear one between the two variables. Although such linear relationships are common, in some cases we are interested in detecting *some* type of relationship between variables, without regard to its specific form. This is often true if the concept in mind does not involve particular variables but rather some more general type of association.

In these cases, rank correlation is often useful. Like the standard correlation coefficient discussed in Section 5.2, the rank correlation coefficient is based on two variables measured for each of $n$ subjects. However, instead of analyzing the numerical values of the variables, we first convert the variables to ranks by computing each subject's rank for each variable. For instance, if $n = 3$ and the pairs of measurements for the 3 subjects are (12,6), (4,9), (8,8), respectively, the respective corresponding ranks are (1,3), (3,1), (2,2) because, for example, 12 is the largest value of the first variable (rank = 1) and 6 is the smallest value of the second variable (rank = 3). If there are ties, then all subjects involved in the tie receive the average of the ranks involved.

Once the data are converted to ranks, we compute the correlation coefficient using the ranks; the result is denoted by $r_S$ and it is often called Spearman's rank correlation coefficient or, simply, the rank correlation coefficient. The rank correlation coefficient has many of the properties of the standard correlation coefficient. For instance, rank correlation always lies between −1 and 1. However, the values 1, −1 no longer indicate a perfect *linear* relationship between the variables, but instead indicate a perfect *monotone* relationship. A monotone relationship is one that is strictly increasing or strictly decreasing. For instance, suppose $X$ is a variable that takes only positive values and suppose $Y = X^2$ exactly. Then, $Y_1 > Y_2$ if and only if $X_1 > X_2$ so that the ranks of the $Y$ values will be exactly the same as the ranks of the $X$ values; in this case, $r_S$ will be exactly 1.

Thus, the properties of the rank correlation coefficient are essentially the same as those of the standard correlation coefficient, except that the linear relationships underlying the standard correlation coefficient are replaced by monotone relationships. For instance, $r_S = 1$ or −1 indicates a perfect monotone relationship between the variables, and $r_S = 0$ indicates that there is no monotone relationship between the variables. Also, the statistical significance of a nonzero value of $r_S$ can be assessed by comparing it to $2/\sqrt{n}$.

# 5.5 THE IMPORTANCE OF A TOP RUNNING BACK IN THE NFL

Suppose we are interested in the importance of a top running back in the offensive production of an NFL team. For each NFL team, we can determine the rank of its most productive running back, in terms of rushing yards, among all team's most productive backs. For example, for the 2011 season, Jacksonville, with Maurice Jones-Drew, receives rank 1 and Detroit receives rank 32. Similarly, we can rank each NFL team in terms of its 2011 regular season scoring so that Green Bay receives rank 1, New Orleans receives rank 2, and so on, down to Saint Louis with rank 32 (Dataset 5.7). The correlation between these sets of ranks is $r_S = -0.050$, indicating essentially no monotone relationship between the sets of ranks.

This result suggests that having a top-level running back is not important in a high-scoring offense. Here, the rank correlation coefficient seems more appropriate than the standard correlation coefficient because we are interested in the general relationship between running back performance and offensive production rather than a specific linear relationship, such as one between rushing yards and points scored.

The low rank correlation between running back performance and offensive production can be contrasted with a similar computation using quarterback ranking, based on the NFL quarterback rating of each team's starting quarterback and team scoring, which yields a rank correlation of 0.810; these data are also available in Dataset 5.7. Therefore, having a top quarterback is much more closely tied to successful offensive production than is having a top running back, a conclusion that is not surprising to anyone who has followed the NFL in recent years. However, it seems a little surprising that having a top running back is apparently completely irrelevant to having a high-scoring offense, at least based on this analysis.

These results can be compared to the results of a similar analysis based on 1975 data (Dataset 5.8). Using those data, the rank correlation coefficient between running back performance and offensive production is 0.599; the rank correlation coefficient between quarterback production and offensive production is 0.815. Therefore, according to this analysis, the relationship between quarterback performance and team scoring was about the same in 2011 as it was in 1975; however, the relationship between a team's top running back and scoring was much stronger in 1975 than in 2011, suggesting an important change in the nature of NFL offenses.

## 5.6 RECOGNIZING AND REMOVING THE EFFECT OF A LURKING VARIABLE

As noted in Section 5.2, an observed correlation between two variables sometimes can be explained by the fact that those two variables are both related to a third variable (said to be a *lurking variable*). In some of these cases, the explanation based on the "third variable" might give a better understanding of the mechanism generating the data than does the original correlation.

For example, consider data on 2009 MLB pitchers with at least 40 innings pitched ($n = 393$) (Dataset 5.9). The correlation coefficient for "hits allowed" and "walks allowed" is 0.77. It is tempting to try to explain this correlation by the theory that pitchers yielding more walks have poor control so they make bad pitches more often, which leads to hits. However, an alternative explanation is that pitchers with a lot of innings pitched naturally give up more hits and more walks; pitchers with fewer innings pitched naturally give up fewer hits and fewer walks. Therefore, the observed high correlation between walks allowed and hits allowed might simply be a consequence of the fact that, for this set of data, innings pitched has considerable variation, and both walks allowed and hits allowed are closely related to innings pitched.

In this example, there is a simple remedy: Instead of analyzing total walks and hits allowed, we could analyze hits allowed per 9 innings pitched and walks allowed per 9 innings pitched. In fact, the correlation coefficient for those variables is 0.0024; that is, there is essentially no linear relationship between hits allowed per 9 innings and walks allowed per 9 innings. Therefore, the explanation that the original high correlation between hits allowed and walks allowed can be attributed to innings pitched seems reasonable and appears to be a better explanation than the one based on control issues.

It is not always possible to use a simple standardization like the one based on innings pitched to control for a third variable. In this section, we present a more general approach to the "lurking variable" effect, as it is sometimes called.

Let $X$, $Y$ denote the variables of interest and let $Z$ denote a third variable that might be at least partly responsible for the linear relationship. Let $r_{xy}$ denote the (standard) correlation coefficient for $X$, $Y$. Let $r_{xz}$ denote the correlation coefficient for $X$, $Z$ and let $r_{yz}$ denote the correlation coefficient for $Y$, $Z$. It may be shown that, if the correlation for $X$, $Y$ is entirely a result of their linear relationships with $Z$, then

$$r_{xy} = r_{xz}r_{yz}.$$

Therefore, $r_{xy} - r_{xz}r_{yz}$ is a measure of the correlation of $X$, $Y$ beyond what can be explained by $Z$.

The *partial correlation coefficient* of $X$, $Y$ controlling for $Z$ is defined by

$$r_{xy \cdot z} = \frac{r_{xy} - r_{xz}r_{yz}}{\sqrt{1 - r_{xz}^2}\sqrt{1 - r_{yz}^2}}.$$

This partial correlation coefficient "controls for $Z$" in the following sense: Suppose that all variable relationships are linear ones, and the relationship between $Y$, $X$ is the same for all values of $Z$. Then, $r_{xy \cdot z}$ measures the correlation between $X$, $Y$ for a sub-population of subjects all having the same value of $Z$, that is, holding $Z$ constant. Stated another way, $r_{xy \cdot z}$ represents an estimate of the correlation between $X$ and $Y$ that would be observed if we were somehow able to observe a sample of subjects all having the same value of $Z$.

Consider the application of these ideas to the example of hits allowed and walks allowed. Let $X$ denote hits allowed, $Y$ denote walks allowed, and $Z$ denote innings pitched. We have seen that $r_{xy} = 0.77$; further analysis shows that $r_{xz} = 0.96$ and $r_{yz} = 0.80$. Note that, to two significant figures, $r_{xy} = r_{xz}r_{yz}$ holds exactly; using exact values shows that $r_{xy \cdot z} = 0.00027$. That is, controlling for innings pitched, there is no linear relationship between hits allowed and walks allowed, essentially the same conclusion we reached using walks and hits per 9 innings.

The assumptions underlying partial correlation are important; hence, it is worth considering them in the context of the example. The first assumption is that the relationships between the variables are all linear ones; this can be addressed by looking at the usual scatterplots of the variables. The second assumption is that the relationship between $X$, $Y$ is the same for all values of $Z$. That is, the relationship between hits allowed and walks allowed is the same for all values of innings pitched. For instance, the relationship is the same for starting pitchers, with 200 or more innings pitched, as it is for closers, with less than 100 innings pitched. Although this assumption is unlikely to be *exactly* true, it does not seem unreasonable; hence, the conclusion that, controlling for innings pitched, there is no relationship between hits allowed and walks allowed is warranted.

Partial correlation coefficients have the same general properties as standard correlation coefficients, except that they measure linear association controlling for the third variable. For example, $r_{xy \cdot z} = 0$ indicates no linear relationship between $X$, $Y$, controlling for $Z$.

It is important to note that the analyst is free to choose the controlling variable that seems appropriate in the analysis; different choices lead to different partial correlations. All such choices are valid, provided that the assumptions discussed previously seem reasonable. These different choices lead to partial correlation coefficients that describe different aspects of the relationship between the variables.

## 5.7 THE RELATIONSHIP BETWEEN EARNED RUN AVERAGE AND LEFT-ON-BASE AVERAGE FOR MLB PITCHERS

Consider the 2009 MLB pitchers' data discussed in the previous section and given in Dataset 5.9. Let $X$ denote the pitcher's earned run average (ERA), and let $Y$ denote his LOBA, left-on-base average (runners left on base per 9 innings). These variables have a correlation coefficient of 0.285, indicating that pitchers with high ERAs tend to leave more runners on base, and conversely, pitchers who leave more runners on base tend to have higher ERAs. To better understand this relationship, we might consider calculating some partial correlations with different control variables. For instance, let B denote walks per 9 innings; then $r_{xy \cdot B} = 0.163$. Let $H$ denote home runs allowed per 9 innings; then, $r_{xy \cdot H} = 0.416$.

Therefore, if we analyze pitchers controlling for walks, the correlation between ERA and LOBA decreases; this might be because pitchers who allow many walks naturally have more base runners. Those who score contribute to ERA; those who do not score contribute to LOBA. The partial correlation for ERA and LOBA controlling for walks is a way to look at the relationship between ERA and LOBA for pitchers with the same number of walks per 9 innings; hence, the effect of walks on the relationship between ERA and LOBA is eliminated, at least approximately. The situation with home runs per 9 innings is more subtle. Home runs increase ERA but decrease LOBA, decreasing the correlation of ERA and LOBA. Restricting attention to pitchers with the same value of $H$ eliminates this effect and, hence, correlation increases. Of course, these explanations are only speculation, but they do suggest avenues for further study.

In some cases, controlling for a variable can have surprising effects. For instance, in the LOBA-ERA example, let B denote WHIP (walks and hits per inning pitched). Then, $r_{xy \cdot w} = -0.998$ That is, controlling for WHIP, LOBA and ERA are almost perfectly negatively correlated. Although this result is, at first, surprising, a little reflection shows why it is to be expected. If we control WHIP, we are controlling the number of base runners. With a few exceptions, each base runner either scores or is left on base. That is, if a runner does not score, he is left on base. Of course, this relationship is not exact because of errors, double plays, and so on, which is why the correlation is "only" −0.998.

# 5.8  USING AUTOCORRELATION TO DETECT PATTERNS IN SPORTS DATA

Correlation measures the strength of the linear relationship between two variables, based on pairs of observations. For instance, if we are interested in the correlation of $X$ and $Y$, we look at pairs of observations of the form $(X, Y)$. Autocorrelation is designed to detect the presence of an approximate linear relationship between adjacent values in a series of measurements on the same variable.

Consider the following example: Let $X$ denote the number of points scored by Chris Paul in a given game. If we observe Paul for the entire 2012–2013 NBA season, we have a sequence of $X$ values, which we can denote by $X_1,\ldots,X_n$, where $n$ denotes the number of observations in the sequence; in this example, $n = 70$ (Dataset 5.10). As discussed in Section 4.10, we often assume that $X_1,\ldots,X_n$ are independent random variables so that the value of one of the $X$ values in the sequence is not related to any of the other values in the sequence. However, suppose that the values are related; for instance, the value of one of the $X$s might be related to the adjacent $X$ values in the sequence. This might be the case, for instance, if Paul tends to "get hot" and have several high-scoring games in a row.

To measure the extent of such a relationship, we might consider the correlation between $X_j$ and $X_{j+1}$, where $j$ is some arbitrary value in the sequence $1,\ldots,n-1$; this correlation is known as the *autocorrelation coefficient* for the sequence. Like a standard correlation coefficient, it is based on pairs of observations, but now the pairs are of the form $(X_j, X_{j+1})$ as $j$ varies from 1 to $n-1$. Note that $j$ must stop at $n-1$, so that $j+1$ cannot exceed $n$. If the values in the sequence are independent, then the "true" autocorrelation coefficient is 0, and the observed autocorrelation coefficient will be close to 0. Note that, because the order of the measurements is important, autocorrelation is only applied to data that are collected in time order. Also, we are implicitly assuming that the correlation between two values in the sequence is the same, no matter where in the sequence they are located; for example, we assume that the correlation between $X_1$ and $X_2$ is the same as the correlation between $X_{n-1}$ and $X_n$.

In the Chris Paul example, the autocorrelation coefficient is −0.018. Because the value is negative, it shows a (slight) tendency for Paul to follow a low-scoring game with a high-scoring one and vice versa. To test the statistical significance of this value, we can compare it to $2/\sqrt{n}$, as we did with a standard correlation. Because here $n = 70$, we have $2/\sqrt{n} = 0.24$, and the observed autocorrelation coefficient is not statistically significant; therefore, it appears that there is little, if any, linear relationship between Paul's points scored in successive games.

Autocorrelation coefficients have the same general properties as other correlation coefficients, except that now the variables in question are the same measurement, but one time period apart. For instance, the autocorrelation coefficient lies between −1 and 1, and it is not affected by changing the units of measurement.

Many sequences of observations are uncorrelated; that is, the autocorrelation coefficient is approximately 0. In these cases, there is little or no linear relationship between adjacent values in the sequence. However, in some cases, adjacent values are related, and if that is the case, taking that relationship into account might be important in the analysis. For instance, consider the season-by-season winning percentages of the NFL teams, starting in 1970, the first season after the merger with the AFL (American Football League). This analysis uses only data from those franchises with a team every season from 1970 to 2012, using data from Pro-Football-Reference.com; therefore, this set of data has results for 25 teams, including the Colts (in Baltimore and Indianapolis), the Rams (in Los Angeles and St. Louis), the Cardinals (in St. Louis and Arizona), and the Oilers/Titans but not the Browns/Ravens.

To estimate the autocorrelation of winning percentages, we can calculate the autocorrelation for each franchise and then average these 25 values; the result is $r = 0.357$. The formula for the margin of error and the method of checking statistical significance discussed previously can be applied here, provided that we make an adjustment to the sample size. Suppose that each autocorrelation is based on $n$ observations for each of $T$ teams; then, use $Tn - 2T$ as the sample size. Note that the total number of observations is $Tn$, so the fact that we are combining results from $T$ teams requires that we decrease the sample size by $2T$. Also note that we are implicitly assuming that the underlying true autocorrelation coefficients for the 25 teams are the same, which is unlikely to be exactly true but is likely to be a reasonable approximation.

Here, $n = 42$ and $T = 25$, so that

$$Tn - 2T = 1000;$$

Because $2/\sqrt{1000} = 0.063$, it follows that the observed autocorrelation is statistically significant.

Therefore, the winning percentage of a team in a given year is related to the winning percentage in the previous year. For instance, if we are interested in predicting a team's winning percentage for the upcoming season, it is important to consider the team's winning percentage last season; the autocorrelation coefficient gives a measure of how closely these values are related, on average.

One way to interpret autocorrelation is to think about predicting $X_{n+1}$, the value in the sequence following $X_n$, based on observing the first $n$ values in the sequence, $X_1, X_2, \ldots, X_n$. If the values in the sequence are independent, then the best predictor is simply the sample mean $\bar{X}$. Now, suppose that the sequence exhibits autocorrelation and let $r_a$ denote the autocorrelation coefficient. Because there is autocorrelation, $X_{n+1}$ is correlated with $X_n$; that is, the value of $X_n$ (already observed) gives us information regarding the as-yet unobserved value of $X_{n+1}$. It follows that the best predictor of $X_{n+1}$ uses a combination of $\bar{X}$ and $X_n$. Specifically, the best predictor is given by

$$r_a X_n + (1 - r_a)\bar{X}.$$

Therefore, if $r_a$ is close to 1, then the predictor is primarily $X_n$; if $r_a$ is close to 0, the predictor is primarily $\bar{X}$.

Now, consider prediction of a team's winning percentage for the 2013 season. Let $W$ denote the team's winning percentage in 2012; here $\bar{X} = 0.500$ (or 50%). Using

$r_a = 0.357$, the predicted winning percentage (written as a proportion, as is commonly done), for the 2013 season is

$$0.357\,W + 0.643\,(.500).$$

For instance, a team winning 5 games in 2012 is predicted to win about 7 games in 2013:

$$0.357\left(\frac{5}{16}\right) + 0.643(0.500) = 0.433 \doteq \frac{7}{16}.$$

Similarly, a team winning 11 games in 2012 is predicted to win about 9 games in 2013.

A more general definition of autocorrelation considers observations that are $h$ units apart in the sequence for various values of $h$. The correlation of observations $h$ units apart is known as the order $-h$ autocorrelation or the *autocorrelation of lag $h$*; in this more general setting, the autocorrelation coefficient described previously in this section is known as the order $-1$ autocorrelation. If the data in the sequence are related, usually we would expect the order $-1$ correlation to be the most important. However, the higher-order autocorrelations give additional information about the relationships in the data.

Consider the NFL winning percentage example. Table 5.1 gives the autocorrelation coefficients for values of the lag $h$ from 1 to 15. Only the values for lags 1, 2, and 3 are statistically significant. Therefore, for small values of $h$, the autocorrelations are positive and decreasing in $h$ until they are essentially zero at $h = 4$. This indicates that the winning percentages for seasons a few years apart are positively correlated, but the correlations decrease quickly as the distance between the seasons increases.

**TABLE 5.1**   Autocorrelations of the NFL winning percentages by year

| LAG | AUTOCORRELATION |
|-----|-----------------|
| 1   | 0.357           |
| 2   | 0.219           |
| 3   | 0.116           |
| 4   | 0.062           |
| 5   | 0.034           |
| 6   | −0.014          |
| 7   | 0.025           |
| 8   | 0.010           |
| 9   | 0.006           |
| 10  | 0.003           |
| 11  | 0.035           |
| 12  | −0.039          |
| 13  | −0.018          |
| 14  | −0.030          |
| 15  | −0.016          |

# 5.9 QUANTIFYING THE EFFECT OF THE NFL SALARY CAP

The autocorrelation coefficients for the NFL winning percentages calculated in the previous section are based on data from the 1970 to 2012 seasons. However, during that time, there was an important change in the NFL: the introduction of the salary cap in 1994. The salary cap is directly relevant to the autocorrelation because one of its intended effects is to increase parity in the league and make it more difficult for a team to have a consistently good, or bad, record. This increased parity, if it exists, should be reflected in the autocorrelation coefficients, which measure the relationship between winning percentages in seasons $h$ years apart, for $h = 1, 2, \ldots$.

To investigate this issue, the autocorrelation coefficients discussed in the previous section were recalculated twice: once using data from the 1970 to 1993 seasons (the precap years) and once using data from the 1994 to 2012 seasons (the postcap years). The same approach used in Section 5.8 was used here; the autocorrelation was calculated for each of the 25 teams, and these results were averaged.

The precap first-order autocorrelation coefficient is 0.364; the postcap value is 0.150. To determine if that difference is statistically significant, we use the approach described in Section 4.7. The first step is to determine the margin of error of each autocorrelation. The precap value is based on 23 observations per team, so that the effective sample size is $(25)(23) - 2(25) = 525$; it follows that the margin of error is

$$2 \frac{\sqrt{1 - (0.364)^2}}{\sqrt{525}} = 0.0813.$$

A similar calculation for the postcap value yields a margin of error of 0.0989. The margin of error of the difference between the two autocorrelation coefficients is therefore

$$\sqrt{(0.0813)^2 + (0.0989)^2} = 0.128.$$

Because the difference between the autocorrelation coefficients is $0.364 - 0.150 = 0.214$, which is about 1.7 times as great as the margin of error, we conclude that the difference between the autocorrelation coefficients is statistically significant; that is, there is evidence that the first-order autocorrelation in winning percentages decreased after the salary cap was adopted.

Further relevant information is provided by the higher-order autocorrelation coefficients. Table 5.2 contains the pre- and postcap autocorrelation coefficients for lags 1 to 5. For the precap coefficients, we judge statistical significance by comparing the magnitude of the coefficient to 0.09; for the postcap coefficients the threshold for statistical significance is 0.10 for the postcap coefficients. Therefore, for the precap era only the coefficients for lags 1 and 2 are statistically significant; for the postcap era, only the lag 1 coefficient is statistically significant.

**TABLE 5.2** Autocorrelation coefficients for NFL winning percentages

| LAG | PRECAP | POSTCAP |
|-----|--------|---------|
| 1 | 0.364 | 0.150 |
| 2 | 0.184 | 0.048 |
| 3 | 0.040 | −0.001 |
| 4 | −0.077 | 0.004 |
| 5 | −0.033 | −0.090 |

Based on these results, we see that in the postcap era there was an important change in the autocorrelation structure of team winning percentages: The first-order autocorrelation is much lower, and the autocorrelations decay to 0 after 1 year, while in the precap era it takes 2 years for the coefficients to become negligible.

# 5.10 MEASURES OF ASSOCIATION FOR CATEGORICAL VARIABLES

The measures of the strength of the relationship between two variables that we have considered so far in this chapter apply to quantitative variables. However, in some cases, the variables of interest are categorical; in this section, we consider methods of measuring the degree of association between such variables. In particular, we focus on the case in which each categorical variable takes two possible values. Similar methods are available for more general categorical variables; however, a number of additional issues arise, making that case beyond the scope of this book.

Consider two categorical variables $X$ and $Y$ and suppose that each variable takes two possible values. For instance, if $X$ denotes the handedness of an MLB pitcher, then $X$ can take the value "L" for left-handed or the value "R" for right-handed. For simplicity, we denote the possible values of each variable as 0 and 1, where the meaning of these values will depend on the variable. For instance, in the example mentioned, we could denote a left-handed pitcher by $X = 0$ and a right-handed pitcher by $X = 1$. Alternatively, we could use $X = 0$ to denote a right-handed pitcher and $X = 1$ to denote a left-handed pitcher. Clearly, the conclusions of the analysis should not depend on how we assign the possible values of $X$ to the values 0 and 1.

Let us look at the following example: For each NFL team's starting quarterback in the 2012 season, consider two variables, one that denotes whether the quarterback was a top 10 pick in the draft and a second that denotes whether the quarterback's team made the playoffs. Here, "starting quarterback" is defined to be the quarterback with the most passing attempts for that team during the regular season; in the case of the San Francisco 49ers, in which both Alex Smith and Colin Kaepernick had 218 attempts, Kaepernick was used as the starter. Therefore, there are 32 data points, corresponding to the 32 starting quarterbacks in the NFL. Table 5.3 contains the data for the example.

**TABLE 5.3**   Draft status and season result for NFL starting quarterbacks in 2012

| | | TEAM MADE THE PLAYOFFS | | |
| --- | --- | --- | --- | --- |
| | | NO | YES | TOTAL |
| Top 10 pick | No | 6 | 5 | 11 |
| | Yes | 14 | 7 | 21 |
| | Total | 20 | 12 | |

To measure the association between these variables, we might consider recoding the data so that the correlation coefficient can be applied. Let $X$ denote the variable representing whether the quarterback was a top 10 pick in the draft and let $Y$ denote the variable representing whether the quarterback's team made the playoffs. To calculate the correlation, we can assign numerical values to "Yes" and "No" for each variable. For instance, let $X = 1$ if the quarterback was a top 10 pick and let $X = 0$ if he was not a top 10 pick; let $Y = 1$ if the quarterback's team made the playoffs and let $Y = 0$ if it did not. The correlation between $X$ and $Y$ is −0.119, indicating that quarterbacks who were top 10 picks are slightly less likely to make the playoffs, or stated another way, teams that made the playoffs are slightly less likely to have a quarterback who was a top 10 pick.

Consider a generic table of data, as given in Table 5.4. Then, coding Yes as 1 and No as 0, an expression for the correlation coefficient is given by

$$r = \frac{ad - bc}{\sqrt{(a+b)(c+d)(a+c)(b+d)}}.$$

It is worth noting that, because the correlation coefficient is unaffected by linear transformations of the variables, the correlation is unaffected by the actual values used to represent the different values of the variable. For instance, if Yes for a top 10 pick is given the value 5 and No is given the value 2 and if Yes for making the playoffs is given the value 10 and No is given the value 6, the correlation is still −0.119. However, if the coding changes the order of the categories, the sign of the correlation may change. For instance, if Yes and No for top 10 pick are given the values 0 and 1, respectively, while Yes and No for making the playoffs are given the values 1 and 0, respectively, then the correlation is 0.119.

**TABLE 5.4**   A generic table representing two categorical variables

| | | VARIABLE 2 | | |
| --- | --- | --- | --- | --- |
| | | NO | YES | TOTAL |
| Variable 1 | No | $a$ | $b$ | $a + b$ |
| | Yes | $c$ | $d$ | $c + d$ |
| | Total | $a + c$ | $b + d$ | |

**TABLE 5.5**   Data on 2012 MLB starting pitchers

|  |  | NO | YES | TOTAL |
|---|---|---|---|---|
|  |  | COMPLETE GAME | | |
| League | National |  |  | 2592 |
|  | American |  |  | 2268 |
|  | Total | 4732 | 128 |  |

The correlation coefficient, as applied to categorical data in this way, retains many of the properties that it has when applied to quantitative data. For instance, recall that a correlation of 0 for quantitative data corresponds to the case of no linear relationship between the variables. In the case of binary categorical variables, an even stronger result holds: A correlation of 0 implies that the variables are independent.

However, there is an important difference between the correlation coefficient for categorical data and the correlation coefficient for continuous data. While, in general, the correlation takes values in the range −1 to 1, when applied to categorical data, the range of the correlation might be restricted by the distributions of the two variables under consideration.

To illustrate this, consider the following example: Suppose we are interested in measuring the association between a starting pitcher's league and the probability that he pitches a complete game. Using data from the 2012 MLB season, there are 4860 starts, 2592 by National League pitchers and 2268 by American League pitchers; there were 128 complete games. Therefore, without knowing how many National League or American League pitchers pitched complete games, the data are of the form given in Table 5.5.

The correlation between league and complete games will take its maximum value if all the complete games are in one league. For instance, if all 128 complete games are thrown by American League pitchers, the data table will be the one given in Table 5.6. The correlation for this table is 0.176; if all the complete games are thrown by National League pitchers, the correlation is −0.154. Therefore, when studying the relationship between complete games and league, the value of the correlation coefficient must fall in the range −0.154 to 0.176. The actual data for this analysis is given in Table 5.7, and the correlation between league and complete game is 0.024, indicating, at most, a weak relationship between league and complete games.

**TABLE 5.6**   Hypothetical data on 2012 MLB starters

|  |  | NO | YES | TOTAL |
|---|---|---|---|---|
|  |  | COMPLETE GAME | | |
| League | National | 2592 | 0 | 2592 |
|  | American | 2140 | 128 | 2268 |
|  | Total | 4732 | 128 |  |

**TABLE 5.7**    Actual data on 2012 MLB starters

|  | | COMPLETE GAME | | |
|---|---|---|---|---|
|  | | NO | YES | TOTAL |
| League | National | 2533 | 59 | 2592 |
| | American | 2199 | 69 | 2268 |
| | Total | 4732 | 128 | |

One reason for this behavior is that the primary interpretation of the correlation as a measure of "how closely the data cluster around a line" does not really apply in the categorical case. Hence, it is often preferable to use a measure of association designed for categorical variables. To do this, it is helpful to think about what it means for two categorical variables to be associated. Consider variables $X$ and $Y$, each of which takes the values 0 and 1. These variables are associated if $Y = 1$ occurs relatively more (or less) frequently when $X = 1$ than it does when $X = 0$.

Note that this is a statement about conditional probabilities or, equivalently, about conditional odds ratios. Let $P(Y = 1 | X = 0)$ denote the conditional probability that $Y = 1$ given that $X = 0$ and let $P(Y = 1 | X = 1)$ denote the conditional probability that $Y = 1$ given that $X = 1$. The odds of $Y = 1$ versus $Y = 0$ when $X = 0$ are given by the ratio

$$P(Y = 1 | X = 0) / P(Y = 0 | X = 0);$$

similarly, the odds of $Y = 1$ versus $Y = 0$ when $X = 1$ are given by the ratio

$$P(Y = 1 | X = 1) / P(Y = 0 | X = 1).$$

Then, $X$ and $Y$ are associated if

$$\frac{P(Y = 1 | X = 0)}{P(Y = 0 | X = 0)} \neq \frac{P(Y = 1 | X = 1)}{P(Y = 0 | X = 1)}.$$

Of course, in practice, we have data, not probabilities. Note that the empirical version of $P(Y = 1 | X = 0)$ is the proportion of times $Y = 1$ occurs, restricting attention to only those cases in which $X = 0$ occurs. For data in the form of Table 5.8, this value is $c/(a+c)$. Similarly, the empirical version of $P(Y = 0 | X = 0)$ is $a/(a+c)$. It follows that the empirical version of the odds ratio $P(Y = 1 | X = 0) / P(Y = 0 | X = 0)$ is $c/a$. Similarly, the empirical version of the odds ratio $P(Y = 1 | X = 1) / P(Y = 0 | X = 1)$ is $d/b$.

To compare the relative likelihoods of $Y = 1$ when $X = 1$ and when $X = 0$, we can look at one odds ratio divided by the other, that is, the ratio of the odds ratios,

$$\frac{P(Y = 1 | X = 1) / P(Y = 0 | X = 1)}{P(Y = 1 | X = 0) / P(Y = 0 | X = 0)}.$$

**TABLE 5.8**   Generic data for two categorical variables

|   |   | X | | |
|---|---|---|---|---|
|   |   | *0* | *1* | *TOTAL* |
| Y | *0* | *a* | *b* | *a + b* |
|   | *1* | *c* | *d* | *c + d* |
|   | Total | *a + c* | *b + d* |   |

This quantity is greater than 1 if $Y = 1$ is relatively more likely when $X = 1$ than when $X = 0$; conversely, it is less than 1 if $Y = 1$ is relatively less likely when $X = 1$ than when $X = 0$. The empirical version of the ratio of odds ratios is

$$\frac{ad}{bc};$$

this is known as the *cross-product ratio* of the data in Table 5.8.

Note that if we switch the roles of $X$ and $Y$ in the analysis, so that we are comparing the likelihood of $X = 1$ when $Y = 1$ to the likelihood of $X = 1$ when $Y = 0$, the empirical version of the ratio of odds ratios is unchanged. Therefore, we can use $\alpha = ad/(bc)$ as a measure of the association between $X$ and $Y$, with a value of $\alpha$ close to 1 indicating a low degree of association. Values of $\alpha$ far from 1, either close to 0 or very large, indicate a high degree of association.

Reciprocals of the cross-product ratio indicate the same degree of association, but in the different direction. For example, values of $\alpha$ of 3 and 1/3 indicate the same degree of association. If, $\alpha = 3$, $Y = 1$ is more likely when $X = 1$; if, $\alpha = 1/3$, $Y = 1$ is more likely when $x = 0$.

For the data in Table 5.3 on the relationship between starting quarterbacks who were a top 10 pick and the team making the playoffs, $\alpha = 0.6$. This means that the odds that a team with a quarterback who was a top 10 pick makes the playoffs are only 0.6 as large as the odds that a team with a quarterback who was not a top 10 pick makes the playoffs. Alternatively, the odds that a playoff team has a quarterback who was a top 10 pick are only 0.6 as large as the odds that a nonplayoff team has a quarterback who was a top 10 pick.

The main advantage of the cross-product ratio as a measure of association is that it is easy to interpret. However, the cross-product ratio does have the disadvantage that its range is 0 to ∞, with 1 indicating no relationship between the variables. Although there is nothing inherently wrong with these properties, it is sometimes easier to work with a measure that has properties similar to those of a correlation coefficient. Let

$$Q = \frac{\alpha - 1}{\alpha + 1}.$$

The quantity $Q$, known as Yule's Q, has the same information as the cross-product ratio but now the range is –1 to 1, with 0 indicating no relationship. Furthermore, like a correlation coefficient, negative and positive values of $Q$ with the same magnitude indicate the same degree of association, but in the opposite direction. For example, for $\alpha = 3$, $Q = 1/2$; for 1/3, $Q = -1/2$. On the other hand, $Q$ does not have the same simple interpretation in terms of relative odds ratios that $\alpha$ does.

For the data in Table 5.3 on the relationship between starting quarterbacks who were a top 10 pick and making the playoffs, $\alpha = 0.6$, so that $Q = -0.25$. It follows that quarterbacks who were top 10 picks are slightly less likely to make the playoffs, the same general conclusion we reached using the correlation coefficient, which is $-0.119$.

One advantage of $\alpha$ and $Q$ over the correlation is that, unlike the correlation, they are not sensitive to the row and column totals. For instance, consider the example on the relationship between complete games and league. For the hypothetical data in Table 5.6, in which all the complete games are thrown by American League pitchers, $\alpha = \infty$ and therefore, $Q = 1$. That is, based on $Q$, there is perfect association between complete games and league, as would be expected using this data; recall that the value of the correlation for this hypothetical data is only 0.176. For the actual data, as given in Table 5.7, $Q = 0.15$.

In general, if the row totals and column totals are greatly different, as in Table 5.7, either $\alpha$ or $Q$ is a more useful measure of the association between the variables than is the correlation. If the row and column totals are roughly similar, as in Table 5.3 (where the row totals are 11 and 21 and the column totals are 12 and 20), then $Q$ and $r$ often show the same general level of association; $\alpha$ still has the advantage of being easier to interpret.

Even if there is actually no association between the variables, we know that the measure of association calculated from data, such as the correlation or $Q$, will not be exactly 0. Therefore, it is often of interest to determine if an observed association is statistically significantly different from 0. First, note that if either the correlation or $Q$ is 0, then the other measure is 0 as well. To see this, note that for the correlation to be 0, we must have

$$ad - bc = 0.$$

If this holds, then $ad = bc$, so that $\alpha = 1$ and, therefore, $Q = 0$. Conversely, $Q = 0$ only if $\alpha = 1$, in which case $ad = bc$, so that the correlation is 0 as well. Therefore, to determine if an observed association is 0, we can compare the correlation $r$ to $2/\sqrt{n}$, as we did when analyzing continuous data.

For instance, for the example on the relationship between starting quarterbacks who were a top 10 pick and making the playoffs, $r = 0.119$ and $n = 32$, so that $2/\sqrt{n} = 0.35$; therefore, the observed association is not statistically significant.

## 5.11  MEASURING THE EFFECT OF PASS RUSH ON BRADY'S PERFORMANCE

In some cases, measures of association for categorical variables are useful even when the underlying variables are continuous if the question being addressed can be expressed in terms of variables that take only two values. For instance, suppose we are interested

**TABLE 5.9** Sacks and touchdown passes for Brady in the 2009–2012 seasons

|  |  | TOUCHDOWN PASSES | | |
| --- | --- | --- | --- | --- |
|  |  | 0–2 | 3 OR MORE | TOTAL |
| Sacks | 0–2 | 29 | 22 | 51 |
|  | 3 or more | 18 | 2 | 20 |
|  | Total | 47 | 24 |  |

in the relationship between Tom Brady's success in passing in a given game and the pressure exerted by the defense's pass rush. Although there are many variables that could be used to address this issue in a formal analysis, here we consider the relationship between touchdown passes, denoted by $T$, and sacks, denoted by $S$.

One approach to measure the extent to which $T$ and $S$ are related is to calculate the correlation for these variables. However, suppose that we are interested in the theory that in games in which Brady is sacked several times he is unlikely to throw several touchdown passes. Although the correlation coefficient tells us something about the degree of linear association between $T$ and $S$, it does not directly address this theory. To do that, we need to look at the relationship between large values of $T$ and large values of $S$. Specifically, let us define "several touchdown passes" to mean three or more and "several sacks" to mean three or more. Therefore, for each game, we record whether Brady threw at least three touchdown passes and whether he was sacked at least three times. The results are given in Table 5.9, using data from the 2009–2012 seasons, including the playoffs.

For the data in Table 5.9, $r = -0.32$. Because $n = 71$, $2/\sqrt{n} = 0.24$; it follows that the observed association is statistically significant. The value of $\alpha$ for these data is 0.15, so that the odds of Brady throwing three or more touchdown passes is about 7 times greater if he is sacked less than three times than if he is sacked three or more times.

# 5.12 WHAT DOES NADAL DO BETTER ON CLAY?

Tennis player Rafael Nadal is considered by some to be the greatest clay-court player of all time, with 9 French Open titles among his 14 grand slam wins (as of June 2014). In this section, we compare Nadal's performance on clay to his performance on other surfaces during the period 2008–2012.

Each game Nadal played during that time period was cross classified in terms of the surface, clay or nonclay, and the result, win or loss; these data were obtained from ATPWorldTour.com. The results are in Table 5.10. Therefore, Nadal won about 66% of the games he played on clay and about 57% of the games he played on other surfaces. The cross-product ratio of the table is 1.43, indicating that the odds that Nadal

**TABLE 5.10**  Nadal's performance on clay and nonclay surfaces

|  |  | RESULT | | |
| --- | --- | --- | --- | --- |
|  |  | WIN | LOSS | TOTAL |
| Surface | Clay | 1660 | 863 | 2523 |
|  | Nonclay | 3658 | 2715 | 6373 |
|  | Total | 5318 | 3578 | 8896 |

wins a game are 1.43 times greater on clay than on nonclay surfaces. The correlation between surface and result is 0.077 with a margin of error of 0.021, which shows that the difference in Nadal's winning percentage between clay and nonclay surfaces is statistically significant.

To further investigate the relationship between the surface and Nadal's performance, we can look separately at games in which he is serving and games in which his opponent is serving. Tables 5.11 and 5.12 contain the results of Nadal's service games and his opponent's service games, respectively, cross classified by surface and result.

The cross-product ratio for games in which Nadal is serving is 1.03, showing that the odds of Nadal winning a service game are roughly the same on clay and on nonclay surfaces. The correlation is 0.0053, which, not surprisingly, is not statistically

**TABLE 5.11**  Nadal's performance on clay and nonclay surfaces on his serves

|  |  | RESULT | | |
| --- | --- | --- | --- | --- |
|  |  | WIN | LOSS | TOTAL |
| Surface | Clay | 1090 | 171 | 1261 |
|  | Nonclay | 2747 | 446 | 3193 |
|  | Total | 3837 | 617 | 4454 |

**TABLE 5.12**  Nadal's performance on clay and nonclay surfaces on his opponent's serves

|  |  | RESULT | | |
| --- | --- | --- | --- | --- |
|  |  | WIN | LOSS | TOTAL |
| Surface | Clay | 570 | 692 | 1262 |
|  | Nonclay | 911 | 2269 | 3180 |
|  | Total | 1481 | 2961 | 4442 |

significant. On the other hand, for games in which Nadal is returning his opponent's serve, the cross-product ratio is 2.05, indicating that his odds of winning such a game are about twice as large on clay as on a nonclay surface. The correlation is 0.158, with a margin of error of 0.030.

Therefore, as expected, Nadal wins a higher percentage of games on clay than on other surfaces. However, his advantage on clay is almost entirely because of his improved performance in the games in which his opponent serves. In the games he serves, there is no evidence that he is better on clay than he is on other surfaces.

# 5.13  A CAUTION ON USING TEAM-LEVEL DATA

It is often convenient to use team-level data to investigate the relationships between variables. It is important to realize that the results of such an analysis apply to teams, but not necessarily to individuals. In particular, it is not generally possible to infer player-level relationships based on team-level data.

Consider the following simple example: Suppose we are interested in the relationship between RBI (runs batted in) and runs scored and suppose we investigate this using team-level data from the 2009–2012 MLB seasons (Dataset 5.11). Using these data, the correlation between RBI per plate appearance and runs scored per plate appearance is 0.995. Such a high correlation is not surprising because, for a given team, each RBI leads to a run scored; of course, not all runs scored are the result of an RBI, so the correlation is not exactly 1.

Based on this result, we conclude that RBI per plate appearance and runs scored per plate appearance are closely related. However, this relationship is only at the team level. The correlation between RBI per plate appearance and runs scored per plate appearance for 2009–2012 MLB players with at least 502 plate appearances is only 0.290, still statistically significant, but not nearly as large as the team-level correlation; these data are available in Dataset 5.12.

In this example, the difference between the team-level and player-level correlations can be attributed to the "teamwork" that is needed to score runs: In most cases, an RBI is needed for the team to score a run. Such a relationship does not hold at the player level.

Team-level and player-level correlations tend to differ under the following conditions. Suppose that we are interested in the relationship between two measurements, $X$ and $Y$. For instance, in the example, we could take $X$ to be RBI per plate appearance and take $Y$ to be runs scored per plate appearance. Team-level and player-level correlations tend to differ whenever the value of $X$ for one player is related to the values of $Y$ for the other players. For instance, in the example, the RBI of one player is related to the runs scored of the other players on the team because often the player's RBI leads to a run scored for another player.

On the other hand, consider the relationship between home runs per plate appearance and doubles per plate appearance for the same players and teams considered in the previous example (Dataset 5.11). Although there might be a slight relationship between the home runs of one player and the doubles of the other players (i.e., both might be because a pitcher is tiring), we expect such a relationship to be relatively weak. The team-level correlation of these statistics is 0.212; the player-level correlation is 0.180. So, although the correlations are not exactly the same, they are fairly close. The margins of error of the team-level and player-level correlations are about 0.18 and 0.08, respectively, so that the difference between them is clearly not statistically significant.

Another reason that team-level and player-level correlations might differ is that there may be much more variability in the player-level measurements than in the team-level measurements, therefore weakening the player-level relationship between the variables. This difference in variability measures often occurs when players have different roles on the team.

Consider the relationship between points per game and three-point shooting percentage in the NBA. Both points per game and three-point shooting percentage are much less variable at the team level than at the player level. For instance, in the 2012–2013 season, team points per game ranged from 93.2 for Washington, Philadelphia, and Chicago to 106.1 for Denver. On the other hand, among players qualified for the scoring title, the points-per-game values ranged from 2.4 to 28.7. This is not surprising because not everyone on a basketball team is expected to be a prolific scorer; on the other hand, all teams have a few players who they rely on to score.

Using data from the 2010–2011 through 2012–2013 seasons (Dataset 5.13), the team-level correlation between points per game and three-point shooting percentage is 0.470, indicating a fairly strong relationship between three-point shooting percentage and scoring. That is, teams with accurate three-point shooting tend to score more than teams that are relatively inaccurate at three-point shots. The player-level correlation, using data in Dataset 5.14, of these variables is only 0.184, indicating that, for players, there is only a weak relationship between scoring and three-point shooting accuracy. This is to be expected because it is not unusual to have a high scoring average without being an accurate three-point shooter.

## 5.14  ARE BATTERS MORE SUCCESSFUL IF THEY SEE MORE PITCHES?

Patience at the plate, as in other aspects of life, is considered to be a virtue. Longer at bats seem to favor the batter and give him more opportunities to get on base; in particular, a walk requires at least 4 pitches. In this section, we consider the relationship between pitches per plate appearance and on-base percentage (OBP) at both the team and player levels.

First, consider team-level data; these data are in Dataset 5.15. Using results from the 2010–2012 seasons, the correlation between pitches per plate appearance and OBP is 0.055. Because there are 90 teams in those seasons, $2/\sqrt{n} = 0.21$; it follows that the correlation is not statistically significant. However, even apart from issues of significance, it is clear that the correlation between pitches per plate appearance and OBP is low at the team level. That is, teams whose batters tend to see more pitches do not have higher OBP than teams whose batters tend to see fewer pitches.

Now, consider player-level data. Using data on players from the 2010–2012 seasons with a qualifying number of plate appearances (Dataset 5.16), the correlation between pitches per plate appearance and OBP is 0.333. There are 443 such players, so that $2/\sqrt{n} = 0.095$; it follows that the correlation is statistically significant. Therefore, batters who see more pitches tend to have higher OBP than batters who see fewer pitches.

Taken together, these results are surprising: Players are more successful if they have more pitches per plate appearance, but teams are not more successful if the players on the team have more pitches per plate appearance. In trying to understand these results, the first step is to examine a plot of the data to make sure that nothing unusual is going on, such as a mistake in the data. These plots are given in Figure 5.7 for the team-level data and Figure 5.8 for the player-level data. In both cases, the value of the correlation coefficient appears to be a reasonable summary of the relationship between the variables. On the team-level plot, there are three data points with the largest values of pitches per plate appearance that are outside the main cloud of points; however, removing these three observations only decreases the correlation, to about 0.03.

There does not appear to be a simple explanation for the different team-level and player-level correlations. In particular, neither OBP nor pitches per plate appearance would seem to be closely related to the pitches per plate appearance or OBP, respectively, of the other players on the team. Also, although there is clearly player-to-player variation in OBP values, this is not by design, as in the case of scoring in the NBA, in which certain players are (informally) designated as "scorers."

One possibility is that, at the team level, the value of pitches per plate appearance is related to some measure of offensive production but not OBP, which is more of a player-level statistic. However, the correlation between pitches per plate appearance and scoring for teams is only 0.126, larger than the correlation between pitches per plate appearance and OBP for teams but still relatively low.

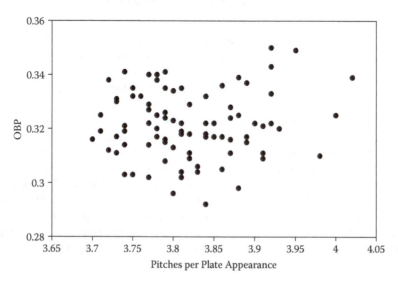

**FIGURE 5.7**    OBP versus pitches per points allowed (PA) for MLB teams.

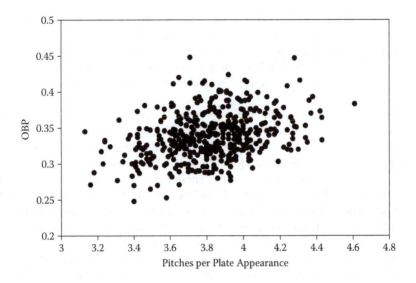

**FIGURE 5.8**   OBP versus pitches per points allowed (PA) for MLB players.

Therefore, we can conclude that players tend to be more successful if they see more pitches, but for teams, offensive success does not appear to be related to the number of pitches seen. However, a good explanation of why this is so is not available.

# 5.15 COMPUTATION

The correlation coefficient can be computed by either using the Correlation procedure in the Data Analysis package or using the CORREL function. Consider the correlation coefficient for runs and hits for 2011 MLB players with a qualifying number of plate appearances. Hence, for each player, we have two variables, runs and hits. The first several rows of this dataset are in Figure 5.9. The entire set of data extends from row 2 to row 146, with row 1 containing the column labels.

The dialog box for the Correlation procedure is given in Figure 5.10. Here, input range refers to a block of cells containing the data, typically organized by columns, but data organized by rows can be accommodated by checking the corresponding box. For example, for the runs and hits example, the input range is B1:C146. It is convenient to include the labels in the input range, in which case the Labels in First Row box must be checked. The output for the procedure is given in Figure 5.11. This should be interpreted as a matrix so that the correlation of R and H is 0.643 while the correlation of either R or H with itself is 1 by definition.

The function CORREL takes two columns of data as its input. For instance, in the example,

*CORREL (B2: B146, C2: C146)*

| | A | B | C |
|---|---|---|---|
| 1 | Name | R | H |
| 2 | Abreu | 54 | 127 |
| 3 | Andino | 63 | 120 |
| 4 | Andrus | 96 | 164 |
| 5 | Avila | 63 | 137 |
| 6 | Aybar | 71 | 155 |
| 7 | Barney | 66 | 146 |
| 8 | Bartlett | 61 | 136 |
| 9 | Bautista | 105 | 155 |
| 10 | Bay | 59 | 109 |
| 11 | Beckham | 60 | 115 |
| 12 | Beltran | 78 | 156 |
| 13 | Beltre | 82 | 144 |
| 14 | Berkman | 90 | 147 |
| 15 | Betancourt | 51 | 140 |

**FIGURE 5.9**  First several rows of the spreadsheet of runs and hits for 2011 MLB players.

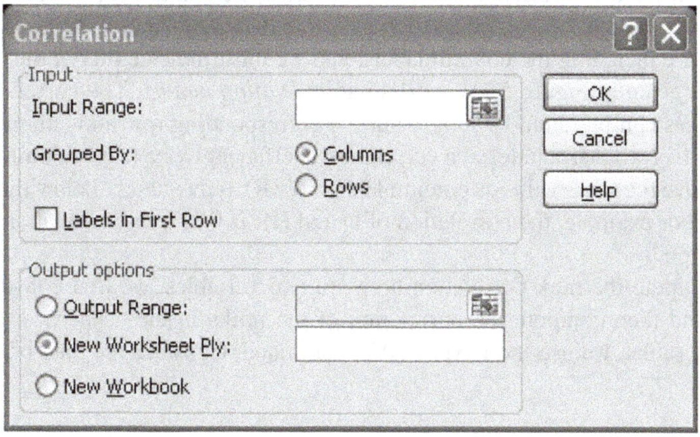

**Correlation**    ? X

Input

Input Range: [                    ]

Grouped By:    ⊙ Columns
               ○ Rows

☐ Labels in First Row

Output options
○ Output Range: [                    ]
⊙ New Worksheet Ply: [                    ]
○ New Workbook

OK
Cancel
Help

**FIGURE 5.10**  The dialog box for the correlation procedure.

| | A | B | C |
|---|---|---|---|
| 1 | | R | H |
| 2 | R | 1 | |
| 3 | H | 0.643231 | 1 |

**FIGURE 5.11**  Output from the correlation procedure.

| | A | B | C | D | E | F | G | H |
|---|---|---|---|---|---|---|---|---|
| 1 | Name | R | H | D | T | HR | SB | BB |
| 2 | Abreu | 54 | 127 | 30 | 1 | 8 | 21 | 78 |
| 3 | Andino | 63 | 120 | 22 | 0 | 5 | 13 | 41 |
| 4 | Andrus | 96 | 164 | 27 | 3 | 5 | 37 | 56 |
| 5 | Avila | 63 | 137 | 33 | 4 | 19 | 3 | 73 |
| 6 | Aybar | 71 | 155 | 33 | 8 | 10 | 30 | 31 |
| 7 | Barney | 66 | 146 | 23 | 6 | 2 | 9 | 22 |
| 8 | Bartlett | 61 | 136 | 22 | 3 | 2 | 23 | 48 |
| 9 | Bautista | 105 | 155 | 24 | 2 | 43 | 9 | 132 |
| 10 | Bay | 59 | 109 | 19 | 1 | 12 | 11 | 56 |

**FIGURE 5.12**  First several rows of the spreadsheet of 2011 MLB batting data.

returns the value 0.643231. One advantage of using CORREL instead of the Correlation procedure is that it is easier to compute the correlation of data in nonadjacent columns.

The Correlation procedure is particularly convenient when computing the correlation of several variables. Consider the 2011 MLB dataset; in addition to runs and hits, the variables doubles ($D$), triples ($T$), home runs ($HR$), stolen bases ($SB$), and walks ($BB$) are also available. Figure 5.12 contains the first several rows of this expanded dataset; the full dataset extends to row 146.

To compute the correlations of these variables, we can use the Correlation procedure but with the input specified as $B1{:}H146$; note that, by including the first row in this range, we are including the column labels, and we must indicate this in the dialog box. The result, given in Figure 5.13, is called a *correlation matrix*. The correlation of any two variables can be found by looking at the corresponding row and column. Because the full matrix would include each correlation coefficient twice (e.g., column R and row H would have the same value as column H and row R), only values "below the diagonal" are given. For example, the correlation of D and HR is 0.317; the correlation of SB and HR is −0.256.

To compute the rank correlation between two variables, we first convert the data to ranks and then compute the correlation of the ranks in the usual way. To convert the data to ranks, we use the RANK.AVG command. Consider the data on the team's

| | A | B | C | D | E | F | G | H |
|---|---|---|---|---|---|---|---|---|
| 1 | | R | H | D | T | HR | SB | BB |
| 2 | R | 1 | | | | | | |
| 3 | H | 0.643231 | 1 | | | | | |
| 4 | D | 0.520792 | 0.638736 | 1 | | | | |
| 5 | T | 0.354261 | 0.220019 | 0.079811 | 1 | | | |
| 6 | HR | 0.494745 | 0.14999 | 0.317066 | -0.25062 | 1 | | |
| 7 | SB | 0.429885 | 0.296721 | 0.014877 | 0.539043 | -0.25571 | 1 | |
| 8 | BB | 0.510778 | 0.173387 | 0.275534 | -0.04849 | 0.476126 | 0.006598 | 1 |

**FIGURE 5.13**  Correlation matrix for the 2011 MLB batting data.

| | A | B | C |
|---|---|---|---|
| 1 | Team | Points | Top Rusher |
| 2 | Arizona | 312 | 1,047 |
| 3 | Atlanta | 402 | 1,340 |
| 4 | Baltimore | 378 | 1,364 |
| 5 | Buffalo | 372 | 934 |
| 6 | Carolina | 406 | 836 |
| 7 | Chicago | 353 | 997 |
| 8 | Cincinnati | 344 | 1,067 |
| 9 | Cleveland | 218 | 587 |
| 10 | Dallas | 369 | 897 |
| 11 | Denver | 309 | 1,199 |
| 12 | Detroit | 474 | 390 |

**FIGURE 5.14**   First several rows of the spreadsheet with the 2011 NFL team data.

most productive running back and the team's scoring for 2011 NFL teams, discussed in Section 5.4. The first several rows of these data are given in Figure 5.14; the full set of data extends to row 33.

To convert Arizona's 312 points to a rank, we use the function

$RANK.AVG(B2,B2:B33),$

which can be interpreted as the function returning the rank of B2 among the rows B2 to B33; this function is entered in cell D2. To be able to copy the function to the entire column, we should use

$RANK.AVG(B2,B\$2:B\$33),$

which keeps the rows B2 to B33 fixed when copying. Copying this function to the entire column yields a column of ranks of team points scored; this is in column D. Similarly, column E contains the ranks of the rushing yards of the team's top rusher. See Figure 5.15 for the first several rows of the spreadsheet that includes the columns of ranks.

The rank correlation is now simply the usual correlation of D2:D33 and E2:E33 calculated using either the Correlation procedure or the CORREL function.

Recall that the (first-order) autocorrelation of a variable $X$, with values $X_1, X_2, \ldots, X_n$ may be viewed as the correlation between $(X_1, X_2, \ldots, X_{n-1})$ and $(X_2, X_3, \ldots, X_n)$. Consider the data on Chris Paul's game-by-game scoring during the 2011–2012 NBA season; suppose that the data are in spreadsheet column A from row 1 to row 60.

One approach to calculating the first-order autocorrelation is to use the command

$CORREL(A1:A59,A2:A60).$

Alternatively, we can construct a second column of data B that is "lagged" by 1 time unit. That is, B1 is A2, B2 is A3, and so on; this is easy to do using the "copy" facility

| | A | B | C | D | E |
|---|---|---|---|---|---|
| 1 | Team | Points | Top Rusher | Pts-Rank | Rush-Rank |
| 2 | Arizona | 312 | 1,047 | 24 | 14.5 |
| 3 | Atlanta | 402 | 1,340 | 7 | 3 |
| 4 | Baltimore | 378 | 1,364 | 12 | 2 |
| 5 | Buffalo | 372 | 934 | 14 | 19 |
| 6 | Carolina | 406 | 836 | 5.5 | 22 |
| 7 | Chicago | 353 | 997 | 17 | 16 |
| 8 | Cincinnati | 344 | 1,067 | 18 | 12 |
| 9 | Cleveland | 218 | 587 | 30 | 30 |
| 10 | Dallas | 369 | 897 | 15 | 21 |
| 11 | Denver | 309 | 1,199 | 25 | 8 |
| 12 | Detroit | 474 | 390 | 4 | 32 |

**FIGURE 5.15**   First several rows of the 2011 NFL team data with ranks.

| | A | B |
|---|---|---|
| 1 | 20 | 10 |
| 2 | 10 | 15 |
| 3 | 15 | 17 |
| 4 | 17 | 20 |
| 5 | 20 | 9 |
| 6 | 9 | 11 |
| 7 | 11 | 27 |
| 8 | 27 | 33 |
| 9 | 33 | 4 |
| 10 | 4 | 18 |
| 11 | 18 | 25 |

**FIGURE 5.16**   Paul's game-by-game scoring lagged by one time period.

in Excel. The result is illustrated in Figure 5.16. The first-order autocorrelation is now just the correlation of columns A and B, keeping in mind that we are using only the first 59 rows so that the input range is A1:B59. Either of these approaches is easily extended to an autocorrelation of an arbitrary order.

# 5.16 SUGGESTIONS FOR FURTHER READING

Correlation is a central topic in statistics and is discussed in (nearly) all books on statistical methods. For further discussion of the correlation coefficient, see the works of Agresti and Finlay (2009, Section 9.4), McClave and Sincich (2006, Section 9.6), and Snedecor and Cochran (1980, Chapter 10). There are many different ways to interpret the correlation coefficient (see Rodgers and Nicewander, 1988). Partial correlation is discussed by Agresti and Finlay (2009, Section 11.7) and Snedecor and Cochran (1980, Section 17.16). See the work of McClave and Sincich (2006, Section 9.10) and Snedecor and Cochran (1980, Chapter 10) for further details on rank correlation.

A sequence of observations measured at successive points in time is known as a time series. Autocorrelation coefficients, as discussed in Section 5.8, are measures of the correlation of the values in a time series. Time series data generally require methods of analysis designed for such data. See, for example, Chatfield's (2003) work for an excellent introduction to the analysis of time series data.

Bishop, Fienberg, and Holland (1975) present a comprehensive introduction to the analysis of the type of cross-classified data; the measures of association presented in Section 5.10 are discussed in Chapter 11 of their work. A more elementary treatment is given by Agresti and Finlay (2009, Chapter 8).

The problem of drawing conclusions about individual behavior, or performance, based on aggregate data, such as team-level data, is known as the ecological-inference problem. It arises in a number of areas of social science, such as political science, when researchers attempt to infer the voting behavior of individuals based on precinct-level data. Robinson (1950) provides a discussion of the relationship between the correlation coefficient based on individuals and the correlation coefficient based on aggregate data; Freedman (1999) contains an excellent summary of the issues involved along with many references. Verducci (2013) discusses the relationship between taking more pitches and success at the plate; the points he raises may be relevant to understanding the results in Section 5.14.

# Modeling Relationships Using Linear Regression

# 6

## 6.1 INTRODUCTION

The correlation coefficient measures the extent to which data cluster around a line. In linear regression analysis, we determine that "best-fitting" line and use it to better understand the relationship between the variables under consideration. The results of linear regression analysis include an equation that relates one variable to another.

Simple linear regression, the subject of this chapter, applies when the data consist of two variables, commonly denoted by $X$ and $Y$, and our goal is to model $Y$, called the *response variable*, in terms of $X$, called the *predictor variable*. Multiple regression is used when our goal is to model a response variable in terms of several predictor variables $X_1, ..., X_p$; models of this type are considered in Chapter 7.

## 6.2 MODELING THE RELATIONSHIP BETWEEN TWO VARIABLES USING SIMPLE LINEAR REGRESSION

Consider the relationship between runs scored in a season and a team's OPS (on-base plus slugging) value for that season for MLB (Major League Baseball) teams from the 2007–2011 seasons (Dataset 6.1). The plot in Figure 6.1 shows that these variables have a strong linear relationship, which is confirmed by the correlation coefficient of 0.96.

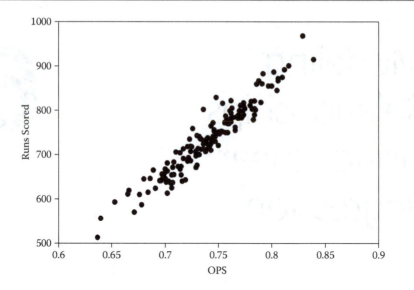

**FIGURE 6.1**   Runs scored versus OPS for 2007–2011 MLB teams.

Let $Y$ denote runs scored and let $X$ denote OPS. The conclusion that $Y$ and $X$ have a linear relationship can be expressed by saying that

$$Y = a + bX$$

for some constants $a$, $b$.

However, it is evident from the plot of runs scored versus OPS that this linear relationship does not hold *exactly*. In statistics, this fact is often expressed by writing

$$Y = a + bX + \varepsilon$$

where $\varepsilon$ represents "random error." Therefore, this equation states that $Y$ is equal to a linear function of $X$ plus random error, or, simply, that $Y$ is approximately a linear function of $X$. It should be noted that the term *error* does not mean "mistake" in this context but rather refers to the deviation from the regression line $a + bX$. An alternative way to write the relationship is as

$$E(Y) = a + b\ X;$$

recall that $E(Y)$ is notation for the average value of $Y$. Therefore, this equation can be interpreted to mean that the average value of $Y$ is equal to $a + bX$. Because the difference between a random variable $Y$ and its average value $E(Y)$ can be described as random error, these two descriptions of the regression model are saying the same thing, and both descriptions are used. In addition, the informal description of the model, $Y = a + bX$, is also sometimes used, with the understanding that the linear relationship between $Y$ and $X$ does not hold exactly.

For instance, consider the example in which $Y$ is runs scored and $X$ is OPS. Here, the linear function of $X$ that best describes $Y$ is $-765 + 2026\ X$; how this function was determined is discussed in the following material. We write the relationship between $Y$ and $X$ as

$$\hat{Y} = -765 + 2026\ X$$

where $\hat{Y}$ represents the value of $Y$ predicted by the regression equation; that is, it represents an estimate of the average value of $Y$ corresponding to $X$.

The regression equation can be interpreted to mean that the average value of runs scored for teams with OPS values of $X$ is $-765 + 2026\ X$. For example, the average number of runs scored for teams with OPS values of 0.750 is

$$-765 + 2026\ (0.750) = 755.$$

However, if a particular team has an OPS of 0.750, we would not expect that team to score *exactly* 755 runs; for example, in 2011, the Brewers had an OPS of 0.750 and scored 721 runs. The difference between the value of $Y$ predicted by the regression equation and the actual observed value can be attributed to the error term ε. Because we are usually concerned with general relationships between variables, rather than specific values that have been observed, we typically ignore the error term, and we refer to it only when it is necessary for understanding the nature of the relationship between $Y$ and $X$.

There are a number of technical conditions that ε must satisfy for the regression model to be useful for describing the relationship between $Y$ and $X$. Generally, these conditions specify that ε corresponds to our intuitive idea of what "random error" should mean. The basic idea is that the data points should be "randomly distributed" around the regression line. For instance, if we consider the entire population of subjects, then ε must have an average value of 0; furthermore, the average value must be 0 if we restrict attention to those subjects with values of $X$ near some specified value. In keeping with the theme of this book, we will not focus on these technical conditions. Instead, we interpret the regression model as stating that $Y$ is approximately a linear function of $X$ with the observations randomly scattered around the line $a + bX$.

The linear regression model states that $E(Y) = a + bX$ for *some* values of $a, b$. When applying this model to data, we choose $a, b$ so that $a + bX$ is the "best-fitting" line to the data, interpreted in the following manner: Consider candidate values for $a, b$, denoted by $a_0, b_0$, respectively. Given a value of $X$, the value of $Y$ corresponding to the line with parameters $a_0, b_0$ is $a_0 + b_0 X$. The difference between the $Y$ value and the point on the line is given by $Y - (a_0 + b_0 X)$, and the squared distance between $Y$ and $a_0 + b_0 X$ is $(Y - a_0 - b_0 X)^2$. We compute this for each pair of $(X, Y)$ data values and sum the results. This gives a measure of how well the line $a_0 + b_0 X$ fits the data; the smaller the value, the better the fit is. We then choose $a_0, b_0$ to make this measure as small as possible; these minimizing values are called the *least-squares* values, and we call the line $a + bX$ the *least-squares regression line* or, simply, the *regression line*.

Although the description of the least-squares approach is a little complicated, $a$ and $b$ are easy to determine numerically for a set of data. For most purposes, it is enough to

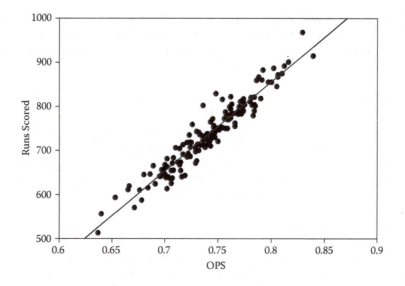

**FIGURE 6.2**   Runs scored versus OPS with regression line.

think of the line $a + bX$ as the line that best fits the data. For the runs scored/OPS example, the estimated regression line is $Y = -765 + 2026\,X$, used previously in this section. Figure 6.2 contains a scatterplot of the data, as in Figure 6.1, but with the regression line added. Clearly, the regression line accurately summarizes the relationship between runs scored and team OPS.

However, we should always keep in mind that "best fitting" is interpreted in terms of distance in the $Y$ direction. In particular, if we switch the roles of $Y$, $X$, we will obtain a different line. For instance, in the runs scored/OPS example, suppose that we take the $Y$ variable to be OPS and we take the $X$ variable to be runs scored. To compare this regression line with the one obtained previously, it is helpful to write both of them in terms of variables $R$ and $P$ where $R$ represents runs and $P$ represents OPS. Then, the "original" regression line based on taking $Y$ to be runs can be written $R = -765 + 2026\,P$. Repeating the analysis with $P$ as the response and $R$ as the predictor leads to the "reversed" regression line $P = 0.408 + 0.000453\,R$. By solving for $R$, the reversed regression line can be written $R = -901 + 2208\,P$, which is similar to, but different from, the original line. Figure 6.3 shows both regression lines on the scatterplot of the data. In general, when comparing the two regression lines in this way, the ratio of the smaller slope to the larger one is $r^2$, where $r$ is the correlation. Therefore, the lines tend to be close to each other when the data closely follow a line, as in this case, in which $r = 0.96$.

A more typical example is given by an analysis relating runs scored to home runs, for MLB teams from the 2007–2011 seasons (Dataset 6.1). Figure 6.4 contains a plot of the data along with both regression lines, one based on taking runs scored as the response variable and home runs as the predictor and one based on taking home runs as the response and runs scored as the predictor. Here, the correlation is 0.69, so the two lines are quite different.

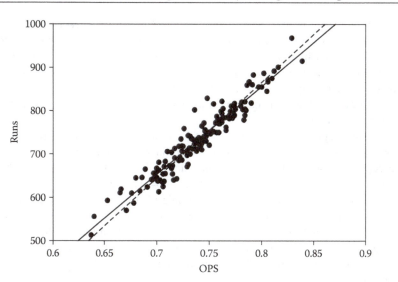

**FIGURE 6.3**   Runs scored versus OPS with two regression lines.

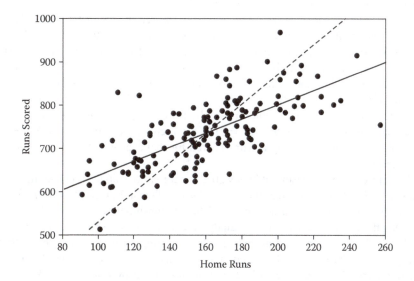

**FIGURE 6.4**   Runs scored versus home runs for 2007–2011 MLB teams.

The lesson here is that, when using linear regression to describe the relationship between two variables, it is important to choose which variable should play the role of $Y$ and which variable should play the role of $X$. When possible, these roles are chosen so that it makes sense to think of $Y$ as "depending on" $X$. For instance, in the example of runs scored versus home runs, it seems to make more sense to think of a team's runs scored as depending on its home runs rather than the other way around.

The linear regression equation $\hat{Y} = a + bX$ relates the variable $Y$ to the variable $X$. The parameters in this relationship, $a$ and $b$, have the usual interpretations as the $y$-intercept and slope, respectively, of the line $a + bX$. That is, $a$ represents the average value of $Y$ corresponding to $X = 0$. Unless $X = 0$ has a particular meaning in the context of the data being analyzed (an example of this is given in Section 6.4), the $y$-intercept is generally not of much interest. The slope $b$ represents the change in the average value of $Y$ corresponding to an increase of 1 unit in $X$. Therefore, in the first example, the estimate $b = 2026$ can be interpreted as meaning that an increase of a team's OPS for a season of 0.100 corresponds to an increase of 202.6 runs for that season.

Recall that the regression equation $\hat{Y} = a + bX$ is a model for the average value of $Y$ corresponding to a particular value of $X$, and the actual observed value of $Y$ differs from this average value by the random error discussed previously. That is, the value of $Y$ is made up of two components: a "systematic" component $a + bX$, which is a linear function of the corresponding value of $X$, and random error. The relative contribution of these components is important for assessing the usefulness of the regression model in describing the relationship between $Y$ and $X$. If the contribution of the random error is small, $Y$ follows the line $a + bX$ closely; if the contribution of the random error is large, $Y$ does not follow the line $a + bX$ closely, if at all.

The $R^2$-$value$ for the regression measures the relative contribution of the systematic component to the values of $Y$. The values of $R^2$ fall between 0% and 100%, with a value of 0% indicating that $Y$ consists entirely of random error, and therefore $a + bX$ is of no help in describing $Y$. A value of 100% indicates that $Y$ is completely determined by $a + bX$, that is, there is no random component in the relationship. The $R^2$-value of a regression is often described as measuring the "percentage of variation in $Y$ explained by $X$" and is sometimes called the *coefficient of determination*. In the example of runs scored/OPS, $R^2 = 91.7\%$, which suggests that a team's OPS for a season is useful in predicting its runs scored for that season, a fact that is apparent from Figure 6.1.

Therefore, $R^2$ is a measure of how well a linear relationship fits the data; hence, we expect $R^2$ to be closely related to the correlation coefficient. In fact, $R^2 = 100r^2$, so that $R^2$ and $r$ measure the same basic aspects of the relationship between $Y$ and $X$.

Using the results of the regression analysis, $\hat{Y} = a + bX$ can be viewed as the "predicted value" of $Y$ corresponding to the value $X$; here, *predicted value* refers to our "best guess" about the value of $Y$ based on knowledge of $X$. The difference $Y - \hat{Y}$ is the "prediction error," called the *residual*, and often denoted by $e$ so that $e = Y - \hat{Y}$. The residual can also be interpreted as the value of $Y$ after the effect of $X$ has been removed. The residuals can be viewed as empirical versions of the "random errors" discussed previously in this section.

# 6.3 THE UNCERTAINTY IN REGRESSION COEFFICIENTS: MARGIN OF ERROR AND STATISTICAL SIGNIFICANCE

Consider a regression model relating a response variable $Y$ to a predictor $X$. The least-squares value of the slope $b$ can be viewed as an estimate of the slope that applies in the "true relationship" between $Y$ and $X$. Hence, the margin of error of an estimate can be used to give a range of values for this true value, as discussed in Chapter 4.

For instance, in the model relating runs scored in a season and a team's OPS, $b = 2026$; the margin of error of this estimate is 100. Therefore, our best guess for the value of the true slope in the regression model with runs scored as the response variable and OPS as the predictor is 2026, and we are reasonably certain that the true value lies in the range 1926 to 2126. Stated another way, based on the data at hand, the value of the slope is 2026; if we were able to observe unlimited data on runs scored and OPS, we are reasonably certain that the value of the slope would be in the range (1926, 2126).

The margin of error is useful for assessing the accuracy of the least-squares estimates and for understanding the range of values that are consistent with the observed data. Note that if $Y$ and $X$ follow a regression model, $Y = a + bX + \varepsilon$ and the true value of the slope is 0, then $Y$ and $X$ are not related. Hence, it is often of interest to determine if the slope estimate is statistically significantly different from 0; in this case, we say simply that the slope is statistically significant. This can be done by comparing the estimate to the margin of error, as discussed in Chapter 4. If the estimate is larger (in magnitude) than the margin of error, then it is statistically significant; otherwise, we cannot exclude the possibility that the true slope is 0. This does not mean that the true slope is *exactly* 0; it just means that, based on our data, there is little or no statistical evidence that the slope is not 0.

In the runs scored/OPS example, the slope estimate is 2026 and the margin of error is 100, so that the slope is clearly statistically significant. Consider an analysis with runs scored as the response variable but with stolen bases, denoted by $S$, as the predictor (Dataset 6.1). The regression equation is

$$\hat{Y} = 712.9 + 0.235\, S.$$

The margin of error of the slope estimate is 0.420. Because 0.235 is less than 0.420, we conclude that the slope estimate is not statistically significant; that is, the data are consistent with a model in which runs scored and stolen bases are not related. Practically speaking, there is not a strong relationship between runs scored and stolen bases, and we would not expect to obtain accurate predictions of a team's runs scored using only stolen bases, a fact that is obvious to most baseball fans.

In determining the statistical significance of a slope estimate, it is important to keep in mind that any conclusions are dependent on the range of predictor variables used in the analysis. For instance, consider NBA (National Basketball Association) centers in the 2010–2011 season who played in at least 70 games (Dataset 6.2). Let $Y$ denote the player's rebounds per 48 minutes played and let $X$ denote the player's height in inches. The regression equation relating $Y$ to $X$ is given by

$$\hat{Y} = 19.1 - 0.075 \ X.$$

The margin of error of the slope estimate is 0.569. It follows that the slope estimate is not statistically significant; hence, there is no evidence of a relationship between height and rebounding. Of course, this does not mean that a player of *any* height can be an effective rebounder in the NBA. All the players in the analysis are between 6'9" and 7'5" tall. What the analysis tells us is that, within that range, for NBA centers playing at least 70 games, height is not a useful predictor of rebounding success.

Another factor affecting the statistical significance of a slope estimate is the sample size. If a slope is not statistically significant, it means that there is insufficient statistical evidence to conclude that it is not 0. This could be because the true slope is 0, or close to 0, or it could be that, although the true slope is not zero, there is insufficient data to detect a relationship between the response and predictor variables.

Consider another analysis of the runs scored of MLB teams, using triples $T$ as the predictor. Using data from just the 2011 season (Dataset 6.1), the estimated regression line is

$$\hat{Y} = 612 + 2.73 \ T.$$

The margin of error of the slope estimate is 3.780, so the estimated slope is not statistically significant. Now, suppose that we repeat the analysis using the data from the 2007–2011 seasons. The estimated regression line is now

$$\hat{Y} = 686 + 1.65 \ T.$$

The margin of error of this slope estimate is 1.508, so the slope estimate is statistically significant. Therefore, teams that hit more triples tend to score more runs, with about 1.65 more runs per triple.

With a larger sample size, the margin of error of an estimate is smaller; hence, more estimates are significant, generally speaking. One consequence of this fact is that, if the sample size is very large, an estimate might be statistically significant but not practically important. For instance, consider MLB pitchers in the 2011 season who pitched at least 36 innings; there are 376 such players. Let $Y$ denote a pitcher's WHIP (walks and hits per inning pitched) and let $A$ denote the pitcher's age in years on June 30, 2011 (Dataset 6.3). The estimated regression equation relating $Y$ to $A$ is given by

$$\hat{Y} = 1.466 - 0.00543 \ A.$$

The margin of error of the slope estimate is 0.00526. It follows that the slope estimate is statistically significant: Older pitchers tend to give up fewer walks and hits per inning pitched.

Note, however, that the magnitude of the effect is quite small. Over 90% of the pitchers studied are between 23 and 36 years old. Using the regression equation, the difference in the average WHIP for 23-year-old pitchers and 36-year-old pitchers is about 0.07, a very small difference. Stated another way, if we are interested in understanding the factors that contribute to a pitcher's WHIP, age is not one that would normally be considered. Therefore, although from a statistical point of view age and WHIP are related, the relationship is not practically important.

# 6.4 THE RELATIONSHIP BETWEEN WINS ABOVE REPLACEMENT AND TEAM WINS

An important contribution of sabermetrics to baseball is the development of methods of measuring the contribution of a player to his team. One of the most useful types of these measures is "wins above replacement" (WAR). Consider the case of a position player. WAR combines the player's contributions in batting, base running, and fielding into a single statistic. The units of WAR are "wins," so that a player with a WAR value of 5, for example, has contributed 5 more wins to his team than a "replacement player" would have. Roughly speaking, a replacement player is the type of player a team might expect to play if a starting player is injured, without expending additional resources (e.g., trading for another team's starting player). Note that, because some positions are easier to play than others, the properties of a replacement player depend on the player's position. For pitchers, WAR is based on the pitcher's contribution to "team defense"; it often uses "fielding-independent" metrics, measures of a pitcher's performance that adjust for the contribution of fielding to pitching statistics.

Many different implementations of this idea have been proposed, leading to several different definitions of WAR (or a similarly named statistic). Here, we use the version of WAR calculated by FanGraphs.com; a detailed description of the calculation used is given on that site. For each player on a team, a value of WAR can be determined, roughly measuring how many wins that player contributed. In 2013, the MLB leader in WAR is Mike Trout, with a value of 10.4. Among pitchers, the leader is Clayton Kershaw, with a value of 6.4. Near the bottom of the WAR list is Paul Konerko, with a value of −1.8, suggesting that the White Sox would have been better off using a generic replacement player in place of Konerko.

One interpretation of the WAR statistic is as a way to distribute a team's wins, above those that would be achieved by a team of replacement players, among the team's players. Because the properties of such a "replacement team" are the same for each MLB team, this suggests that a team's actual wins should be closely related to its WAR, the sum of the WAR values for each player on the team. More specifically, if $W$ represents a team's actual wins and $X$ denotes the team's WAR, we expect that $W \doteq R + X$, where $R$ represents the number of wins expected from a replacement team; note that it is generally thought that $R$ is about 50.

To investigate this relationship, we can conduct a linear regression analysis with team wins as the response variable and team WAR as the predictor variable. Using data from the 2009–2013 MLB seasons, the estimated regression equation is

$$\hat{W} = 49.7 + 0.940\, X.$$

The margin of error of the intercept estimate is about 2.8, so that the results are consistent with the value of wins by a replacement team in the range 47 to 52.5. The margin of error of the slope estimate is 0.08, so that the results are consistent with a slope of 1, as expected based on the theory of WAR. However, because 1 is near the end of the range $0.940 \pm 0.08$, the results suggest that WAR might be slightly overestimating the number of wins attributed to each player.

The value of $R^2$ for the regression is 78.1%, indicating that about 78% of the variation in team wins can be explained by the team's WAR values; that is, team WAR is a good predictor of team wins. To evaluate the magnitude of this $R^2$ value, we can compute the $R^2$ value for the regression of team wins on other possible predictors. For instance, if we use the difference of a team's hits and walks and its hits and walks allowed as the predictor, we obtain $R^2 = 71.7\%$. Consider a team's total bases plus its hits and walks, similar to what would be used in calculating its OPS. If we use the difference between that value and the total bases plus hits and walks that a team has allowed as the predictor, then $R^2 = 79.0\%$, slightly higher than what we obtained using team WAR as the predictor. Therefore, although team WAR is a useful predictor of team wins, it is not necessarily better than other measures of team performance.

This result is not surprising because the ability of a team to hit well and to prevent its opponent from hitting well is generally considered to be the most important factor in a team's success. Note that this result does not suggest that WAR should be replaced by a simpler measure, such as total bases plus hits and walks. The primary motivation of WAR is not to predict team wins but to evaluate individual players. The results presented here are designed to check if WAR is calibrated correctly (it appears to be) and if it is closely related to team wins, as we would expect (it is).

## 6.5  REGRESSION TO THE MEAN: WHY THE BEST TEND TO GET WORSE AND THE WORST TEND TO GET BETTER

A regression model is often used as a way to model the relationship between the variables. Another use for a regression model is to better understand how one variable is related to the other. One of the most important results of this type is a property known as *regression to the mean*.

Consider variables $Y$ and $X$ that are related through a linear regression model. Suppose that we convert each variable to its $Z$-score so that $Y$ is converted to

$(Y - \bar{Y})/S_y$, where $\bar{Y}$ denotes the average value of the $Y$ values and $S_y$ is the standard deviation of the $Y$ values; similarly, suppose that $X$ is converted to $(X - \bar{X})/S_x$. If we do the regression analysis in terms of the $Z$-scores, it can be shown that the $y$-intercept is always 0 and the slope is exactly $r$, the correlation coefficient. Therefore, the relationship between $\hat{Y}$ and $X$ may be written as

$$\frac{\hat{Y} - \bar{Y}}{S_y} = r \frac{X - \bar{X}}{S_x}$$

When interpreting this equation, keep in mind that $\hat{Y}$ represents an estimate of the average value of $Y$ corresponding to $X$. Because $|r| \leq 1$,

$$\frac{|\hat{Y} - \bar{Y}|}{S_y} \leq \frac{|X - \bar{X}|}{S_X}$$

and, unless $r$ is 1 or $-1$,

$$\frac{|\hat{Y} - \bar{Y}|}{S_y} < \frac{|X - \bar{X}|}{S_X}$$

That is, $\hat{Y}$ is closer to $\bar{Y}$ than $X$ is to $\bar{X}$, where "closer" is measured relative to the standard deviation of the variables. This phenomenon is known as regression to the mean; it says that, if $Y$ and $X$ are linearly related, then, unless it is a perfect linear relationship (with a correlation of 1 or $-1$), the $Y$ values corresponding to a given value of $X$ will tend to be closer to the average $Y$ value than $X$ is to the average $X$ value, again measuring closeness relative to the standard deviations.

Consider the following example. For 2009 MLB players with a qualifying number of plate appearances, let $Y$ denote the player's value of weighted on-base average (WOBA) and let $X$ denote the player's batting average (BA); the data used here are taken from BaseballGuru.com. As would be expected, BA and WOBA are related, with a correlation coefficient of 0.58. In 2009, the top 5 MLB players in BA were Joe Mauer, Ichiro Suzuki, Hanley Ramirez, Derek Jeter, and Pablo Sandoval; these 5 players had an average BA of.345. The MLB-wide average BA was .282 with a standard deviation of 0.0257. Thus, the top 5 average BA of .345 is 2.45 standard deviations above the league average BA because

$$\frac{.345 - .282}{.0257} = 2.45.$$

Now, consider the WOBA values of these players. Because WOBA and BA are related, we expect them to have relatively high values of WOBA. However, because of regression to the mean, we expect their WOBA values to be less extreme than their BA values. That is, in fact, the case. The five players who led the league in BA have an average WOBA of .403; the league average of WOBA is .353 with a standard deviation

**TABLE 6.1**   Some examples of regression to the mean

| NBA GUARDS TOP 10 IN FT PCT IN 2010–2011 | | | NHL PLAYERS TOP 10 IN SHOTS IN 2010–2011 | | |
|---|---|---|---|---|---|
| | RANK IN FT PCT | RANK IN 3PT PCT | | RANK IN SHOTS | RANK IN GOALS |
| Curry | 1 | 2 | Ovechkin | 1 | 14 |
| Billups | 2 | 19 | Byfuglien | 2 | 97 |
| Nash | 3 | 27 | Carter | 3 | 7 |
| Augustin | 4 | 82 | Kessel | 4 | 14 |
| Meeks | 5 | 24 | Zetterberg | 5 | 54 |
| Foye | 6 | 77 | Nash | 6 | 14 |
| Martin | 7 | 34 | Gionta | 7 | 30 |
| Allen | 8 | 1 | Staal | 8 | 12 |
| Paul | 9 | 31 | Perry | 9 | 1 |
| Collison | 10 | 75 | Iginla | 10 | 3 |

of 0.0352. Therefore, the average WOBA value for these 5 players is only 1.42 standard deviations greater than the league average.

One way to interpret regression to the mean is to note that if a player is highly ranked according to one variable, then he will tend to be less highly ranked according to a second variable. Table 6.1 contains some examples of this type. Consider the data on the top 10 NBA guards in free-throw (FT) percentage. Clearly, the ability to shoot free throws is related to the player's ability to make three-point (3PT) shots. However, the relationship is not a perfect one. Therefore, we know that a player's 3PT shooting percentages will be less extreme than the player's FT shooting percentages, on average. This is confirmed by the data in Table 6.1. Of the top 10 guards in FT percentage, only 2 are in the top 10 in 3PT percentage.

Note that regression to the mean is a statement about averages, not individual observations. Therefore, it is possible for a given player to have extreme values of both variables; an example of this is Stephen Curry, who ranked first in FT percentage and second in 3PT percentage. However, on average, highly ranked FT shooters are less proficient at 3PT shooting. Similar considerations apply to the hockey data in Table 6.1. Players who are highly ranked in terms of the number of shots taken tend to be less highly ranked in terms of goals scored. However, there are some players who are highly ranked in terms of both variables.

If the variables $X$, $Y$ are such that $S_y \doteq S_x$, then the regression effect is more straightforward: It says simply that subjects with above-average (or below-average) values of $X$ will tend to have values of $Y$ that are more average. Cases for which $S_y \doteq S_x$ often occur when $Y$ and $X$ represent measurements of the same variable in two different years. For example, Table 6.2 lists the 5 NFL (National Football League) running backs with the most rushing yards in 2009 together with their 2010 rushing yardage. Note that all 5 gained less yardage in 2010 than in 2009. Similarly, the 2009 NFL sack leaders all had fewer sacks in 2010 than in 2009.

**TABLE 6.2**  Further examples of regression to the mean

| | 2009 NFL RUSHING LEADERS | | | 2009 NFL SACK LEADERS | |
|---|---|---|---|---|---|
| | 2009 RUSHING YARDS | 2010 RUSHING YARDS | | 2009 SACKS | 2010 SACKS |
| Johnson | 2006 | 1364 | Dumervil | 17 | 0 |
| Jackson | 1416 | 1241 | Allen | 14.5 | 11 |
| Jones | 1402 | 896 | Freeney | 13.5 | 10 |
| Jones-Drew | 1391 | 1324 | Woodley | 13.5 | 10 |
| Peterson | 1383 | 1298 | Smith | 13 | 5.5 |
| | 2010 NFL RUSHING LEADERS | | | 2010 NFL SACK LEADERS | |
| | 2010 RUSHING YARDS | 2009 RUSHING YARDS | | 2010 SACKS | 2009 SACKS |
| Foster | 1616 | 257 | Ware | 15.5 | 11 |
| Charles | 1467 | 1120 | Hali | 14.5 | 8.5 |
| Turner | 1371 | 871 | Wake | 14 | 5.5 |
| Johnson | 1364 | 2006 | Matthews | 13.5 | 10 |
| Jones-Drew | 1324 | 1391 | Abraham | 13 | 5.5 |

Note that this "regression effect" has nothing to do with the statistics themselves or with the fact that the variables are ordered in time; it is a consequence of the approximate linear relationship between the variables. In fact, it works "in reverse" as well. For instance, in Table 6.2, the top NFL rushers in 2010 tended to have less yardage in 2009 (although 2 of them actually had more yardage in 2009—remember regression to the mean applies to averages, not individual measurements), and the sack leaders in 2010 all had fewer sacks in 2009.

Although the regression effect is common and occurs in all fields, it is often mistaken for something more substantive; such an erroneous analysis is sometimes said to be based on the *regression fallacy*. For example, consider the following "theory" on detecting when a top MLB batter is beginning to decline. According to this theory, if an older player, who is a top performer, has several times been hit by a pitch (HBP) in a given season, then this suggests that his reflexes are degrading; hence, his performance will soon suffer—call this the "HBP theory." To test the HBP theory, I identified players who in 2010 had an OPS value of .850 or above; who were 30 years old or older on June 30, 2010; and who had 7 or more HBP (the median value for MLB is 4). There are 5 such players: Jose Bautista, Matt Holliday, Jayson Werth, Aubrey Huff, and Ryan Howard. Four of those 5 players had a lower OPS value in 2011 than in 2010, with an average change in OPS of −.078 for the five players, which might be viewed as providing strong support for the HBP theory.

Although the HBP theory *might* be correct, these results are more likely to be an example of the regression fallacy. OPS values greater than .850 are above average;

hence, those batters on my list can be expected to have an OPS value closer to the league average in 2011. To investigate this possibility, I analyzed the players with an OPS of .850 or greater in 2010 who were 30 years old or older on June 30, 2010, but who did not meet my HBP criterion. There are 7 such players. Of these 7, there were 6 who had a lower OPS value in 2011, with an average change in OPS for the group of −.067. These results strongly suggest that the HBP theory is not correct, and the observed results are simply a consequence of the regression effect.

One lesson here is that, because of regression to the mean, it is rarely a good idea to base conclusions of an analysis on only the top or bottom players (or teams).

# 6.6 TRYING TO DETECT CLUTCH HITTING

Many baseball analysts are suspicious of the idea of "clutch hitting," believing instead that any difference in hitting in clutch and nonclutch situations is caused by random variation in the data. See, for example, the work of Silver (2006b) and Tango, Lichtman, and Dolphin (2007, Chapter 4).

To investigate this issue, I performed the following analysis: FanGraphs (http://www.FanGraphs.com) rates each batter's plate appearance as low, medium, or high "leverage." A high-leverage appearance is one that has a relatively large effect on the outcome of the game.

To measure a player's "clutchness," I computed the difference between his OPS in high-leverage appearances and his OPS in low-leverage appearances; call this "clutch-OPS." Therefore, a large positive value of clutch-OPS means a player performs better in high-leverage situations than in low-leverage situations, measuring performance by OPS. Using data from the 2011 and 2012 MLB seasons, I computed the clutch-OPS for those players with at least 50 high-leverage plate appearances in both seasons ($n = 84$); note that roughly 10% of all plate appearances are high-leverage ones. Table 6.3 contains the top and bottom 10 MLB players in terms of 2011 clutch-OPS.

To determine if clutch hitting is genuine, the following argument is sometimes used. If clutch hitting actually exists, the best clutch hitters in 1 year should also be good clutch hitters in other years. Conversely, if clutch hitting is caused by randomness, the best clutch hitters in 1 year will not necessarily be good clutch hitters in other years. The same general considerations apply to the worst clutch hitters.

Table 6.3 also contains the clutch-OPS in 2012 for the top and bottom clutch hitters in 2011. Note that the top clutch hitters in 2011 did not do as well in 2012, with 8 of 10 showing a decrease in clutch-OPS; the average decrease was 0.183. Similarly, all 10 of the bottom 10 clutch hitters in 2011 showed an improvement in clutch-OPS in 2012, with an average increase of 0.304. Results like these are sometimes used to suggest that clutch hitting is, in fact, caused by randomness rather than some skill that players possess.

However, we know that, because of regression to the mean, the best players in terms of any 2011 statistic will always tend to have lower values of that statistic in 2012; similarly, the worst players in terms of any 2011 statistic will always tend to have a higher

**TABLE 6.3**   Top and bottom 10 clutch-OPS for 2011 MLB players

|  | TOP 10 IN 2011 | | | BOTTOM 10 IN 2011 | |
| --- | --- | --- | --- | --- | --- |
|  | 2011 | 2012 |  | 2011 | 2012 |
| Votto | .448 | .567 | Suzuki | −.463 | .107 |
| Smoak | .379 | −.293 | Escobar | −.386 | −.001 |
| Pena | .377 | .166 | Kendrick | −.357 | −.334 |
| Descalso | .320 | −.069 | Ruiz | −.326 | .151 |
| Fielder | .316 | .249 | Ibanez | −.323 | .166 |
| Johnson | .296 | .043 | Rollins | −.267 | .179 |
| Willingham | .271 | .506 | Stanton | −.253 | .017 |
| Headley | .267 | .128 | Lee | −.238 | .068 |
| Cano | .253 | −.164 | Francoeur | −.233 | −.231 |
| Kemp | .251 | .217 | Beckham | −.226 | −.153 |
| Average | .318 | .135 | Average | −.307 | −.003 |

value of that statistic in 2012. For instance, consider the OPS value itself. Table 6.4 contains the top and bottom 10 players in 2011 OPS along with their 2012 OPS for all MLB players with a qualifying number of plate appearances in both years ($n = 98$).

Note that all players in the top 10 in OPS in 2011 showed a decrease in OPS in 2012, with an average decrease of 0.063. Eight of the bottom 10 in OPS in 2011 showed an increase in 2012, with an average increase of 0.096. In comparing the changes in average values in Tables 6.3 and 6.4, it is important to keep in mind that OPS values generally have a smaller range than clutch-OPS values because clutch-OPS is computed as the difference between two OPS measurements. Therefore, the OPS values

**TABLE 6.4**   Top and bottom 10 OPS for 2011 MLB players

|  | TOP 10 IN 2011 | | | BOTTOM 10 IN 2011 | |
| --- | --- | --- | --- | --- | --- |
|  | 2011 | 2012 |  | 2011 | 2012 |
| Cabrera | 1.034 | .999 | Rios | .613 | .850 |
| Braun | .994 | .987 | Escobar | .633 | .721 |
| Fielder | .981 | .940 | Beckham | .633 | .668 |
| Gonzalez | .958 | .806 | Gonzalez | .642 | .760 |
| Ortiz | .952 | .859 | Suzuki | .645 | .696 |
| Granderson | .916 | .811 | Hill | .655 | .882 |
| Holliday | .913 | .877 | Desmond | .656 | .845 |
| Beltran | .910 | .842 | Barney | .666 | .653 |
| Pujols | .907 | .859 | Stubbs | .685 | .610 |
| Konerko | .905 | .857 | Prado | .687 | .796 |
| Average | .947 | .884 | Average | .652 | .748 |

in Table 6.4 have the same general properties as the clutch-OPS values in Table 6.3. It follows that if we conclude that clutch-OPS is essentially random based on Table 6.3, we should conclude that OPS is essentially random based on Table 6.4.

The problem with both analyses is that it is difficult to account for the impact of regression to the mean in the data in Tables 6.3 and 6.4. A better way to approach the clutch hitting issue is to analyze *all* the players, not just those with highest or lowest values of clutch-OPS. For instance, the correlation between clutch-OPS in 2011 and clutch-OPS in 2012 is 0.027, which is not statistically significant. This suggests that, in fact, clutch hitting can be attributed to randomness rather than to a genuine skill. For comparison, the correlation between OPS in 2011 and OPS in 2012 is 0.509, which is statistically significant. Therefore, analyses based on the entire set of data show an important difference between OPS and clutch-OPS. There is a linear relationship between a player's OPS in different years so that players with a high (or low) value of OPS in 1 year tend to have a high (or low) value of OPS in the next year; on the other hand, there is no such relationship between a player's clutch-OPS value in 1 year and the next.

# 6.7  DO NFL COACHES EXPIRE?
# A CASE OF MISSING DATA

Although it is often misleading to analyze only the players or teams with the highest (or lowest) value of some statistic, in some cases such an analysis cannot be avoided. This is particularly true when analyzing sports data because only the best players receive playing time, only the best teams make the playoffs, and so on. Consider the following example: It is sometimes suggested that NFL coaches have a "shelf life" of approximately 10 years, after which they become less successful (Joyner, 2008, Chapter 7). According to Pro-Football-Reference.com, at the conclusion of the 2012 NFL season, 51 coaches had coached for more than 10 years. To investigate the "10-year shelf-life" theory, we can compare the winning percentage of each of these coaches during their first 10 years to their winning percentage for seasons coached after this initial 10-year period. For 38 of the 51 coaches, their winning percentage in the 11th and later seasons was lower than their winning percentage in the first 10 seasons; for these 51 coaches, the average decrease in the proportion of games won is 0.102. Therefore, this analysis seems to support the 10-year shelf-life theory.

However, note that for a coach to be included in the study, he must have coached at least 10 years; a coach with a low winning percentage in his first several years is unlikely to coach 10 years. In fact, only about 11% of NFL coaches have coached more than 10 years. We know that, because of regression to the mean, coaches with a high winning percentage in their first 10 years will tend to have a lower winning percentage in their 11th and later seasons. If this is a result of regression to the mean, we would expect coaches with a below-average winning percentage during their first few years to have a higher winning percentage in later years. Unfortunately, after a few poor seasons, coaches often do not have further opportunities to coach, making a full analysis impossible. Therefore, we can speculate that regression to the mean may be playing a role here, but more definitive proof is not available.

## 6.8 USING POLYNOMIAL REGRESSION TO MODEL NONLINEAR RELATIONSHIPS

Consider a simple linear regression model relating a response variable $Y$ to a single predictor $X$:

$$\hat{Y} = a + bX + \varepsilon$$

Although this model is often useful, in some cases a line is not an accurate description of the relationship between $Y$ and $X$. In these cases, a higher-degree polynomial is often useful for modeling the relationship.

Consider the following example in which we analyze Ken Griffey Jr.'s home run frequency over the course of his career (Dataset 6.4). For the response variable $Y$, take his home runs per 100 at bats ($HR$) in a given season, and for the predictor variable we use the year of his career, denoted by $T$, so that $T = 1$ denotes his rookie year. Figure 6.5 contains a plot of these data from Griffey's first 21 major league seasons (his brief final season is not included), together with the regression line, using a linear relationship $\hat{Y} = a + bT$.

Clearly, the regression line does not adequately describe the relationship between $HR$ and $T$. For instance, note that the points with either small or large values of $T$ tend to lie below the line, while points with values of $T$ near the center of the plot tend to lie

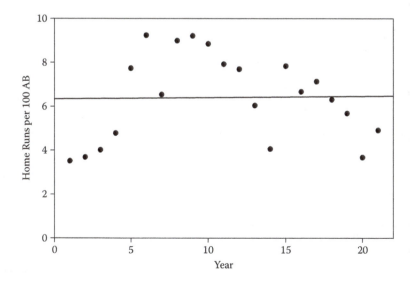

**FIGURE 6.5**   Griffey's home run rate by year.

above the line. That is, the error term ε in the regression model tends to be negative for small and large $T$ and positive otherwise. This contradicts the assumption that the error term has mean 0 for any values of the predictor variable. More important, a line misrepresents the nature of the relationship between $HR$ and $T$. It suggests that as Griffey ages, his home run rate will stay approximately constant; based on this model, there was no reason for him to retire!

One way to deal with such a nonlinear relationship is to use a *polynomial regression model*. In a polynomial regression model, the linear function $a + bX$ is replaced by a polynomial in $X$ of the form

$$a + b_1\, X + b_2\, X^2 + \cdots + b_p\, X^p$$

where $a$, $b_1$, ..., $b_p$ are parameters to be estimated. The simplest polynomial regression model is the quadratic model

$$Y = a + b_1\, X + b_2\, X^2 + \varepsilon.$$

The shape of the quadratic function $a + b_1\, X + b_2\, X^2$ is a parabola, with axis of symmetry

$$X = -b_1/2b_2.$$

If $b_2 > 0$, the minimum of the quadratic occurs at $-b_1/2b_2$; if $b_2 < 0$, the maximum occurs at $-b_1 / 2b_2$. See Figure 6.6 for some examples of quadratic functions.

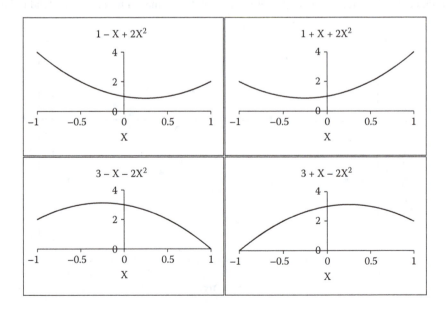

**FIGURE 6.6**   Some examples of quadratic functions.

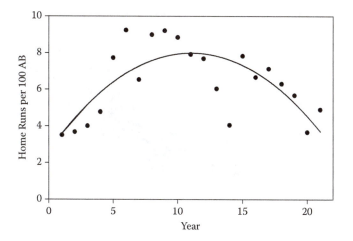

**FIGURE 6.7**   Griffey's home run rate by year with quadratic regression curve.

Figure 6.7 contains a plot of Griffey's home run data, together with a quadratic regression function. The equation for this function is

$$\hat{Y} = 2.69 + 0.957\ T - 0.0433\ T^2;$$

it takes its maximum value at

$$T = -\frac{0.957}{2(-0.0433)} = 11.1$$

so that, according to this model, Griffey's home run frequency increases through his first 11 seasons and then begins to decrease.

Although polynomial regression models are most useful when, as in the Griffey example, the relationship between the response variable and the predictor is clearly nonlinear, even in cases in which the relationship is approximately linear, including a quadratic term might improve the performance of the model. Consider the following example: Playing time for NHL (National Hockey League) players generally varies considerably, with top players playing 20 or more minutes per game and some role players playing 10 or fewer minutes. Naturally, players who play more often score more. Let $Y$ denote the points per game of a given player over a season, and let $T$ denote that player's average time on ice (TOI) for that season. Using data from the 2011–2012 season for all forwards who played at least 60 games and averaged at least 6 minutes per game (Dataset 6.5), a regression analysis, using a linear function, was performed. The resulting regression equation is

$$\hat{Y} = -0.505 + 0.0639\ T,$$

with an $R^2$ value of 71.9% (see Figure 6.8).

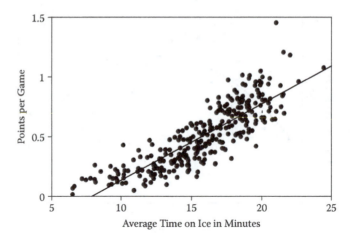

**FIGURE 6.8**   Points per game versus average TOI for NHL forwards.

Note, however, that the plot of the data together with the regression line given in Figure 6.8 suggests that there is a nonlinear component to the relationship; for instance, the points tend to lie above the regression line for small and large values of TOI. Using a quadratic regression model leads to the regression equation

$$\hat{Y} = 0.0905 - 0.0201\ T + 0.00280\ T^2,$$

with an $R^2$ value of 74.4%. More important, the plot of the data together with this model, given in Figure 6.9, shows that the quadratic regression model does a better job of capturing the relationship between points scored and TOI.

Furthermore, the quadratic model has important implications for the relationship between points per game and playing time. For instance, according to the linear model, increasing playing time by 1 minute per game corresponds to an increase in scoring of approximately 0.064 points per game, or about 5 points over an 80-game season. According to the quadratic model, increasing playing time from 10 minutes to 11 minutes corresponds to an increase of approximately 3.1 points over an 80-game season, while increasing playing time from 20 minutes to 21 minutes corresponds to an increase of approximately 7.6 points over an 80-game season.

In interpreting results of this type, it is important to keep in mind that the relationship between points per game and TOI is not a causal one. For instance, if the playing time of St. Louis right wing Ryan Reaves is increased from the roughly 6.5 minutes per game he played in 2011–2012 to 20 minutes a game, we would not expect his scoring to increase by 0.73 points per game, as suggested by the quadratic regression equation—there are reasons why Reaves only played 6.5 minutes per game (e.g., he had only 4 points in 60 games in 2011–2012).

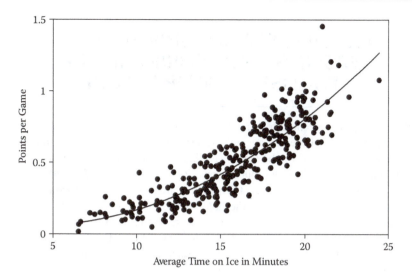

**FIGURE 6.9**  Points per game versus average TOI for NHL forwards.

Note that a linear relationship between variables $X$ and $Y$ is a special case of a quadratic relationship because, if $b_2 = 0$ in the function $a + b_1 X + b_2 X^2$, then we are left with the linear function $a + b_1 X$. Therefore, we can determine if the quadratic term in the relationship is needed by considering the statistical significance of $b_2$. In the NHL forwards example, $b_2 = 0.00280$ and the margin of error is roughly 0.001. Therefore, $b_2$ is statistically significant, and the quadratic model is an improvement over the linear one.

For the example in which $Y$ is runs scored and $X$ is OPS, considered in Section 6.2, if we fit a quadratic model, the coefficient of the quadratic term in $X$ is 1604 with a margin of error of 1869. Because the margin of error is greater than the magnitude of the coefficient itself, we conclude that the coefficient of the quadratic term is not statistically significant; hence, we conclude that the linear model is adequate for describing the relationship between runs scored and OPS.

Returning to the NHL forwards example, note that the coefficient of $b_1$ in the quadratic model is $-0.0201$, with a margin of error of 0.032. Therefore, the coefficient is less in magnitude than the margin of error. Although $b_1$ is not statistically significant in this case, it is generally not a good practice to remove it from the model if the quadratic term is still present. The reason is that a quadratic model without a linear term is a specific type of function; for example, the function $a + b_2 X^2$ is symmetric about $X = 0$ with minimum value at $X = 0$. On the other hand, such a function is only marginally simpler than the full quadratic model, $a + b_1 X + b_2 X^2$. Therefore, in most cases, setting $b_1 = 0$ imposes a strong condition on the relationship between $X$ and $Y$ without simplifying it in any useful way.

# 6.9 THE RELATIONSHIP BETWEEN PASSING AND SCORING IN THE ENGLISH PREMIER LEAGUE

Clearly, passing plays a central role in soccer. In this section, we consider the relationship between passing, as measured by a team's pass success rate, and scoring, as measured by the goals per game scored by the team. The data consist of the season results from the English Premier League (EPL) for four seasons, 2009–2010 through 2012–2013; therefore, there are 80 observations in the dataset. These data are taken from WhoScored.com.

Figure 6.10 contains a plot of the goals per game versus the pass success rate. Although the correlation is fairly high (0.65), clearly the relationship is nonlinear, being essentially constant for pass success rates in the range of 62% to 75% and then curving upward from 75% to 85%. Therefore, this is a situation in which it may be useful to use a polynomial regression model to describe the relationship.

Let $G$ denote the goals per game of a given team for a particular season, and let $P$ denote the corresponding value of the pass success rate. Using a quadratic regression model of the form

$$G = b_0 + b_1 P + b_2 P^2 + \varepsilon,$$

the estimated regression model is

$$\hat{G} = 12.0 - 0.329\, P + 0.00247\, P^2;$$

the value of $R^2$ is 48.8%. Because the margin of error of $b_2$ is 0.00163, it is statistically significantly different from 0; hence, the quadratic relationship is an improvement over a linear one, in which $b_2 = 0$.

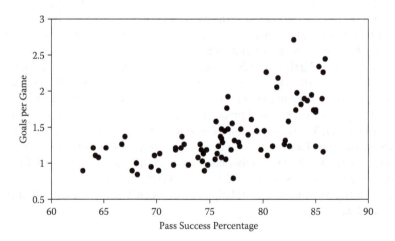

**FIGURE 6.10**   Scoring versus pass success percentage in the EPL.

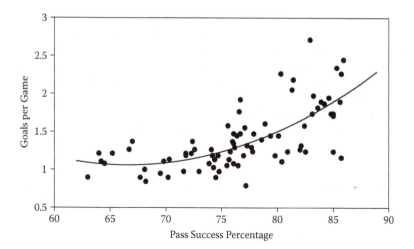

**FIGURE 6.11**   Scoring versus pass success percentage in the EPL.

Figure 6.11 contains a plot of the data together with the estimated regression function. Clearly, the quadratic model does a fairly good job of capturing the relationship, particularly in the center of the range of pass success rate. For very low values of pass success rate, the function begins rising; such behavior is a consequence of using a quadratic function, which is symmetric about its axis of symmetry. Also, the right hand of the estimated function does not appear to rise as quickly as the data suggest. Again, this is a consequence of using a quadratic function: If the function were to rise more steeply on the right, it would also need to rise more steeply on the left, making it a poor fit to the data. It is possible to try a cubic function in place of the quadratic; however, the cubic term is not significantly different from 0, and the fit is essentially the same.

Therefore, according to the estimated quadratic regression function, for teams that pass poorly, the relationship between pass success rate and scoring is nearly constant. For instance, the estimated difference in scoring corresponding to changing the pass success rate from 65% to 70% is only about 0.02 goals per game. However, the situation is quite different for teams that are more successful at passing. For instance, the estimated difference in scoring corresponding to changing the pass success rate from 85% to 90% is about 0.5 goals per game, an important difference.

# 6.10 MODELS FOR VARIABLES WITH A MULTIPLICATIVE EFFECT ON PERFORMANCE USING LOG TRANSFORMATIONS

Suppose we are interested in the relationship between a response variable $Z$ and a predictor variable $W$. A regression model of the form $E(Z) = a + bW$ has the property that an increase in $W$ of 1 unit corresponds to a change of $b$ units in the average value of $Z$.

That is, a change in $W$ of a fixed size corresponds to a change in the average value $Z$ of fixed size. However, in some cases, a change in $W$ of fixed size corresponds to a *proportional* change in the expected value of $Z$. In other cases, a proportional change in $W$ corresponds to a fixed change in the average value of $Z$. These types of relationships can be modeled through the use of a *log transformation*. Such transformations were discussed in Section 2.8 in the context of a single variable. Here, we consider the use of log transformations in modeling the relationship between variables.

Recall that the natural log of $a$, $\ln(a)$, is defined by the relationship

$$e^{\ln(a)} = a$$

where $e$ is a number known as the base of the natural logarithms, and $\ln(a)$ can only be calculated if $a$ is a positive number. Also, if $a$ and $b$ are positive numbers, then

$$\ln(ab) = \ln(a) + \ln(b)$$

and, for any $c$,

$$\ln(a^c) = c\ln(a),$$

so that the ln function transforms multiplication into addition, in a certain sense. Therefore, use of the log transformation changes the nature of the relationship between the response and predictor variables.

Consider the NHL example in Section 6.8. Let $P$ denote the points per game of a player and let $T$ denote his average TOI. Previously, we modeled the relationship between $P$ and $T$ using a quadratic regression model with $Y = P$ as the response variable. An alternative approach is to consider transformations of either one or both of $P$ and $T$.

For instance, consider using $Y = \ln(P)$ as the response variable in a simple linear regression model. Figure 6.12 contains a plot of $Y$ versus $T$; note that the relationship appears to be linear. The estimated regression equation is

$$\hat{Y} = -3.39 + 0.160 \ T.$$

Figure 6.12 gives a plot of the transformed points-per-game data versus average TOI. Such plots are useful because it is easy to see if a linear relationship between the transformed variables is appropriate. However, it is also useful to see a plot of the data and regression function using the original data. Using the defining property of the ln function, we can write the linear relationship between $\ln(P)$ and $T$ in terms of the original variable $P$:

$$\hat{P} = 0.0337 \ e^{0.160T}.$$

A plot of this relationship, together with the original data, is given in Figure 6.13. Note that the regression function fits the data well.

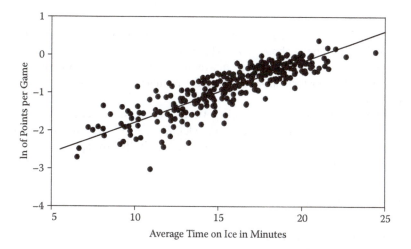

**FIGURE 6.12**   ln(PPG) (log of points per game) versus average TOI in the NHL.

An important property of a regression model based on a transformation is what it says about the relationship between the original variables. Using the equation relating the estimated average points per game $\hat{P}$ to time on ice $T$, we see that increasing $T$ by 1 unit changes $\hat{P}$ from $0.0337\,e^{0.160T}$ to $0.0337\,e^{0.160(T+1)}$. Because

$$0.0337\,e^{0.160(T+1)} = 0.0337\,e^{0.160T}\,e^{0.160}$$

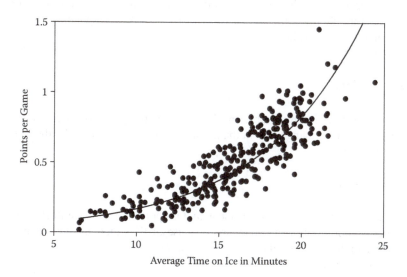

**FIGURE 6.13**   Points per game versus average TOI in the NHL.

and $e^{0.160} = 1.174$, increasing $T$ by 1 minute per game corresponds to an increase in the average points per game of about 17.4%. That is, a unit increase in the predictor variable corresponds to a *percentage* increase in average value of the response variable. This property of the ln transformation is one of the primary reasons that it is so useful in modeling relationships.

In general, consider the relationship between a response variable $Z$ and a predictor variable $W$. Using the standard regression model, $Z = a + bW + \varepsilon$, we have seen that an increase of $W$ of 1 unit corresponds to a change of $b$ in the average value of $Z$. Now, consider the model

$$\ln(Z) = a + b\,W + \varepsilon;$$

we may think of this as a model with $Y = \ln(Z)$ as the response variable and $W$ as the predictor. Then, using the defining property of the ln function,

$$Z = e^{a+b\,W+\varepsilon}.$$

Using properties of exponents, this implies that

$$Z = e^{a+\varepsilon}\, e^{b\,W};$$

because the first term in this expression does not depend on $W$, we can interpret this as meaning that the average value of $Z$ is of the form $c\,e^{b\,W}$ for some constant $c$ not depending on $W$. Therefore, increasing $W$ by 1 unit corresponds to a *proportional increase* in the average value of $Z$ of $e^b$. It follows that using the ln transformation for $Z$ might be appropriate if we think that a 1-unit increase in $W$ corresponds to a *percentage* increase in the average value of $Z$.

Now, consider the model

$$Z = a + b\ln(W) + \varepsilon$$

in which $Z$ is the response variable and $X = \ln(W)$ is the predictor. Increasing $X$ by 1 unit corresponds to a change of $b$ in the average value of $Z$. Because, by the properties of the ln function,

$$1 + \ln(W) = \ln(e\,W)$$

an increase in $\ln(W)$ of 1 unit corresponds to a proportional increase in $W$. Therefore, this model is appropriate if a percentage increase in $W$ is thought to correspond to a fixed change in the average value of the response variable $Z$.

Similarly, the model

$$\ln(Z) = a + b\ln(W) + \varepsilon,$$

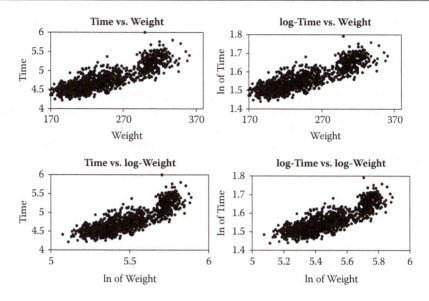

**FIGURE 6.14** Plots of 40-yard dash time versus weight.

which is a simple linear regression model for $Y = \ln(Z)$ and $X = \ln(W)$, is useful when a percentage increase in $W$ is thought to correspond to a percentage change in $Z$.

Therefore, in many cases, assumptions about the nature of the relationship between two variables can suggest which transformation, if any, to use. For instance, consider the NFL prospects who participated in the NFL draft combine during the period 2008–2012 and suppose we are interested in the relationship between a prospect's weight and his time in the 40-yard dash; the data analyzed here are available on NFLCombineResults.com. Let $W$ denote a player's weight in pounds and let $T$ denote the player's time in seconds in the 40-yard dash. Figure 6.14 contains a plot of different combinations of log-transformed data for the 1321 players who recorded an official 40-yard dash time. Note that the plots do not give much information regarding possible transformation; for each of the four cases considered, there appears to be a strong linear relationship between the variables.

To see the effect of log-transforming $T$ or $W$ or both, we can look at the various estimated regression lines. With both variables in their original form, the estimated regression equation is

$$\hat{T} = 3.30 + 0.00606 \ W;$$

according to this equation, a 10-pound increase in weight of a player corresponds to about a 0.06-second increase in his 40-yard dash time. Note that because regression relationships are not causal, "increasing weight" does not refer to adding weight to a physically fit player; it refers to studying physically fit players weighing more than the first group of players considered.

Using $Y = \ln(T)$ and $W$ as the predictor yields the estimated regression line

$$\widehat{\ln(T)} = 1.26 + 0.00124 \ W.$$

Using this equation, a 10-pound increase in the weight of a player corresponds to an increase of 0.012 in the log of the player's 40-yard dash time. Because $e^{0.012} = 1.012$, this can be interpreted as saying that a 10-pound increase in the weight of a player corresponds to an increase of about 1.2% in the player's 40-yard dash time. Note that for small values of $x$, $e^x$ is approximately $1 + x$.

Using $Y = T$ and $X = \ln(W)$ as the predictor yields the estimated regression line

$$\hat{T} = -3.39 + 1.49\ln(W).$$

Suppose that we increase $W$ by 5% so that it changes from $W$ to $1.05W$. Then, the log of $W$ changes from $\ln(W)$ to

$$\ln(1.05 \ W) = 0.049 + \ln(W);$$

that is, $\ln(W)$ increases by about 0.049. This corresponds to an increase in 40-yard dash time of $1.49(0.049) = 0.073$; that is, a 5% increase in weight corresponds to an increase of about 0.07 seconds in 40-yard dash time.

Finally, the estimated regression line for $Y = \ln(T)$ and $X = \ln(W)$ is

$$\widehat{\ln(T)} = -0.118 + 0.306\ln(W).$$

According to this equation, a 5% increase in weight corresponds to roughly a 1.5% increase in 40-yard dash time. This result is obtained as follows: Suppose $W$ is increased by 5%; let $W_1 = 1.05 \ W$ denote the new weight. Then,

$$\ln(W_1) = \ln(1.05 \ W) = \ln(1.05) + \ln(W) = 0.0488 + \ln(W)$$

so that $\ln(W)$ increases by 0.0488. It follows that $\ln(T)$ increases by

$$(0.306)(0.0488) = 0.0149$$

and therefore $T$ increases by

$$e^{0.0149} = 1.0150,$$

about 1.5%.

When summarizing the relationship between weight and 40-yard dash time, we need to choose which of these four analyses to use. In some cases, plots of the transformed data suggest the best choice; as noted previously, that is not true here. Formal measures of goodness of fit, such as $R^2$, are of limited usefulness. In this example, the regression of $T$ on $W$ has $R^2 = 74.8\%$, the regression of $\ln(T)$ on $W$ has $R^2 = 75.1\%$, the regression of $T$ on $\ln(W)$ has $R^2 = 72.8\%$, and the regression of $\ln(T)$ on $\ln(W)$ has $R^2 = 73.3\%$. Note that although it might be useful to use $R^2$ to compare the two models with response variable $T$ or the two models with response variable $\ln(T)$, $R^2$ cannot be used to compare models with different response variables because the models are trying to predict different quantities. Therefore, we can conclude that $W$ is a slightly better predictor of either $T$ or $\ln(T)$ than is $\ln(W)$, although the $R^2$ values are so close that this conclusion is not definitive.

In cases such as this one, it is often better to think about which type of relationship seems most appropriate for the data being analyzed. Note that the equations of elementary physics relating the time required to move an object of weight $W$ a fixed distance using a fixed force suggest that $\ln(T)$ is a linear function of $\ln(W)$ with slope 0.5. Although here we do not expect the force to be fixed—a bigger player generally generates more force—the idea that $\ln(T)$ should be approximately a linear function of $\ln(W)$ seems valid. Therefore, it is reasonable to summarize these data by saying that a 5% increase in weight corresponds to a 1.5% increase in 40-yard dash time.

The regression equation relating $\ln(T)$ and $\ln(W)$ can also be used to identify those players who are particularly fast for their weight. Let $W_0$ denote the weight of a given player and let $T_0$ denote his time in the 40-yard dash. According to the regression equation we have derived, the average of the log of the 40-yard dash time for a player with weight $W_0$ is

$$-0.118 + 0.306\ln(W_0).$$

The difference between the player's actual log-time and the average log-time for players of his weight is

$$\ln(T_0) - [-0.118 + 0.306\ln(W_0)].$$

This quantity, which can be described as the observed response minus the average response predicted by the regression equation, is known as a residual. In this example, players with negative residuals are fast relative to players of their weight.

For the 1321 players analyzed in the combine example, the player with the most negative residual is Dontay Moch, who weighed 248 pounds and ran a 40-yard dash in 4.40 seconds. The list of those in the top 14 in this measure (roughly the top 1%) includes 4 Pro-Bowl players (Von Miller, Jimmy Graham, Trent Williams, and Patrick Peterson).

So far in this section, we have used transformations as part of the process of finding a simple relationship between variables. Another role for transformations is to convert a particular nonlinear relationship that we wish to use into a linear one that can be handled by linear regression methodology.

For instance, consider the relationship implied by the "Pythagorean formula" used in baseball, as well as in other sports; this was discussed in Section 5.3. Let $W$ and $L$ denote a team's wins and losses, respectively, and let $F$ and $A$ denote the team's "points for" and "points against," respectively. According to the Pythagorean relationship, there is a constant $\beta$ such that

$$\frac{W}{W+L} = \frac{F^\beta}{F^\beta + A^\beta} .$$

Note that this relationship can be rewritten as

$$\frac{1}{1+L/W} = \frac{1}{1+A^\beta / F^\beta}$$

and, using a little algebra, as

$$\frac{W}{L} = \left( \frac{F}{A} \right)^\beta .$$

Taking the ln transformation of both sides leads to the relationship

$$\ln\frac{W}{L} = \beta \ln\frac{F}{A} .$$

Let $Y = \ln(W/L)$ and $X = \ln(F/A)$. Then, $Y$ and $X$ follow a linear regression model with slope $\beta$ and intercept equal to 0.

Using data from the 2008–2012 MLB seasons (Dataset 6.6), a regression model with response variable $\ln(W/L)$ and predictor variable taken to be the natural log of the ratio of runs scored to runs allowed yields an estimate of the Pythagorean exponent of 1.79, with an $R^2$ value of 87.6%. The intercept in this model is estimated to be 0.00104, which is close to, but not exactly, 0. If we force the intercept to be exactly 0, the estimate of the exponent is unchanged at 1.79. This is in close agreement with the value 1.83 that is commonly used.

# 6.11  AN ISSUE TO BE AWARE OF WHEN USING MULTIYEAR DATA

In the linear regression model

$$Y = a + bX + \varepsilon,$$

the term $\varepsilon$ represents "random error," so $Y$ is randomly distributed around the line $a + bX$. That is, we assume that there is no discernible pattern to the values of $\varepsilon$.

Suppose we are studying the relationship between two variables by analyzing several years of data on teams or players. For concreteness, consider the example in Section 6.4 in which we studied the relationship between a team's actual wins $W$ and its total WAR value $X$ for its players. In fitting a regression model with response variable $W$ and predictor $X$, we used 5 years of data, from the 2009 MLB season to the 2013 season. Therefore, each team appears 5 times in the dataset, once for each of these 5 seasons.

For instance, consider the 2009 Phillies. Their WAR value for that season is 42.2; using the regression equation,

$$\hat{W} = 49.7 + 0.940\ X,$$

the predicted value of $W$ corresponding to $X = 42.2$ is

$$49.7 + 0.940\ (42.2) = 89.4.$$

In 2009, the Phillies actually won 93 games so that their value of $W$ for that year is 93. That is, using the regression equation, they won more games than would be predicted from their value of WAR.

The difference between the actual value of $W$ (93) and the value predicted from the regression equation (89.4) is 3.6 wins. This difference between the actual and predicted values is known as a *residual*. A residual can be viewed as an estimate of the error term $\varepsilon$.

The fact that the 2009 Phillies won more games than predicted by their value of WAR might be because of random luck, but it is more likely caused by a combination of luck and the fact that the team has some properties that are not captured by WAR but that contribute to their success. In fact, in 2010 they also outperformed the value predicted by WAR; the value of the residual for the Phillies in 2010 is 9.7 wins. For the remaining 3 years considered in this analysis, the residuals for the Phillies are 8.7, −5.5, and 7.6, respectively. Therefore, in 4 of the 5 years considered, the Phillies won more games than predicted by their WAR values.

This suggests that the error term in the regression equation is not completely random because errors for the same teams over several seasons are related. Of course, teams change from year to year, so we would not expect a strong relationship among the error terms for a given team, but it would not be surprising if there is *some* relationship. In the team wins/WAR example, the correlation between residuals from the same team is 0.088, so the residuals are slightly correlated. It is important to note that this correlation does not apply to the values of $W$ themselves; that correlation is much higher, 0.34 for these data. The correlation 0.088 refers to the correlation in the differences between the actual and predicted wins for the same team over several seasons. Also, note that the methods needed to calculate these correlations require statistical methods that are beyond the scope of those covered in this book.

This type of situation is not uncommon whenever the dataset consists of players or teams measured several times over the course of several seasons. In those cases, we do not expect the assumption that the error terms are completely random to be literally true. Fortunately, even if this assumption is not satisfied, the estimates obtained in the linear regression analysis are still valid descriptions of the relationships in the data.

However, there are two ways in which the analysis might be affected. One is that we might be able to determine more accurate estimates if we use an analysis that recognizes the fact that the same teams or players are included in the dataset several times. Such an analysis requires advanced statistical methods, as well as more sophisticated software, and those methods are not discussed here; see Section 6.13 for references. Another issue is that the margins of error obtained from the regression output may be inaccurate; however, they are typically still useful as general guides when the correlation between residuals is relatively small, as in the team wins/WAR example.

For instance, in the WAR and wins example, if we fit a model that allows for correlation between the error terms of the same teams in different years, the estimated regression equation is

$$\hat{W} = 49.5 + 0.944\ X;$$

as noted previously, obtaining these results requires specialized statistical software. This regression equation is close to, but not exactly the same as, the one obtained in Section 6.4,

$$\hat{W} = 49.7 + 0.940\ X.$$

The margin of error for the slope parameter in the more sophisticated analysis is 0.086, which is slightly larger than the margin of error obtained originally, 0.082. However, the difference is unlikely to be important when using the results.

# 6.12 COMPUTATION

Calculations for linear regression are performed using the Regression procedure in the Data Analysis package. Consider the example on the runs scored and OPS by year of MLB teams from the 2007–2011 seasons. The first several rows of the spreadsheet are shown in Figure 6.15; the entire set of data extends to row 151.

The dialog box for the Regression procedure is given in Figure 6.16. "Input Y Range" refers to the location of the response variable in the regression, and "Input X Range" refers to the location of the predictor variable. Therefore, in the runs-OPS example, these are A1:A151 and B1:B151, respectively; because I have included the column labels in these ranges, the "Labels" box must be checked.

The output for this regression analysis is given in Figure 6.17. The regression coefficients are given in the table at the bottom of the output. "Intercept" refers to $a$; OPS refers to the coefficient of OPS in the regression model. Therefore, the regression line is

$$Y = -765 + 2026X$$

| | A | B | |
|---|---|---|---|
| 1 | Runs | OPS | |
| 2 | 855 | 0.8 | |
| 3 | 875 | 0.81 | |
| 4 | 787 | 0.773 | |
| 5 | 730 | 0.744 | |
| 6 | 762 | 0.766 | |
| 7 | 718 | 0.725 | |
| 8 | 867 | 0.788 | |
| 9 | 721 | 0.75 | |
| 10 | 735 | 0.739 | |
| 11 | 615 | 0.684 | |

**FIGURE 6.15**  First several rows of the runs-OPS data from 2007–2011 MLB teams.

where $Y$ denotes runs and $X$ denotes OPS. The standard error of the coefficient of OPS is given in the column adjacent to the coefficient column; recall that the margin of error is simply twice the standard error, so the margin of error for the OPS coefficient is about 100. The "t Stat" column in this table refers to the coefficient divided by its standard error, known as the $t$-statistic. Therefore, if the $t$-statistic is less than 2 in absolute value, the coefficient is not statistically significantly different from 0. The value of $R^2$

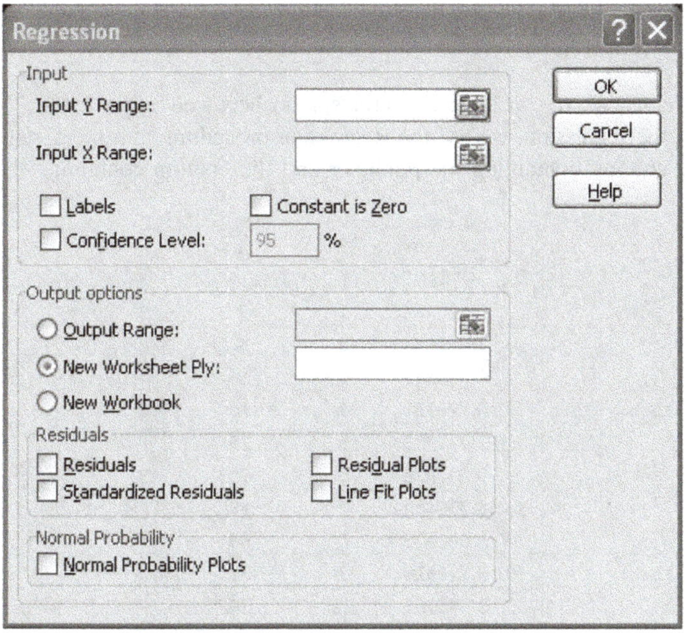

**FIGURE 6.16**  The dialog box for the regression procedure.

| | A | B | C | D | E | F | G | H | I |
|---|---|---|---|---|---|---|---|---|---|
| 1 | SUMMARY OUTPUT | | | | | | | | |
| 2 | | | | | | | | | |
| 3 | Regression Statistics | | | | | | | | |
| 4 | Multiple R | 0.95774494 | | | | | | | |
| 5 | R Square | 0.91727538 | | | | | | | |
| 6 | Adjusted R Sq | 0.91671643 | | | | | | | |
| 7 | Standard Error | 22.8962279 | | | | | | | |
| 8 | Observations | 150 | | | | | | | |
| 9 | | | | | | | | | |
| 10 | ANOVA | | | | | | | | |
| 11 | | df | SS | MS | F | Significance F | | | |
| 12 | Regression | 1 | 860309.1267 | 860309.127 | 1641.06828 | 5.4896E-82 | | | |
| 13 | Residual | 148 | 77587.11329 | 524.237252 | | | | | |
| 14 | Total | 149 | 937896.24 | | | | | | |
| 15 | | | | | | | | | |
| 16 | | Coefficients | Standard Error | t Stat | P-value | Lower 95% | Upper 95% | Lower 95.0% | Upper 95.0% |
| 17 | Intercept | -764.6958 | 37.09902239 | -20.61229 | 2.4176E-45 | -838.00802 | -691.38359 | -838.00802 | -691.38359 |
| 18 | OPS | 2025.86354 | 50.00884942 | 40.510101 | 5.4896E-82 | 1927.03993 | 2124.68715 | 1927.03993 | 2124.68715 |

**FIGURE 6.17**   Regression output for the runs-OPS example.

is given in the table at the top of the output, as a proportion not a percentage; we can simply multiply the reported value by 100 to convert it to a percentage. For example, here $R^2 = 91.7\%$.

To fit a quadratic regression model, we need a second predictor variable representing the squared term. Consider the example on Griffey's home run rate by year. As in Section 6.7, let $Y$ denote his home run rate and let $T$ denote the year, ranging from $T = 1$ for his rookie year to $T = 21$. To fit the model

$$Y = a + b_1 T + b_2 T^2 + \epsilon$$

we need to construct the variable $T^2$ in the spreadsheet (see Figure 6.18).

To run the regression, we use the Regression procedure with the $X$ range taken to include the column containing the predictor and the column containing the predictor

| | A | B | C |
|---|---|---|---|
| 1 | HR Rate | T | T-squared |
| 2 | 3.516484 | 1 | 1 |
| 3 | 3.685092 | 2 | 4 |
| 4 | 4.014599 | 3 | 9 |
| 5 | 4.778761 | 4 | 16 |
| 6 | 7.731959 | 5 | 25 |
| 7 | 9.237875 | 6 | 36 |
| 8 | 6.538462 | 7 | 49 |
| 9 | 8.990826 | 8 | 64 |
| 10 | 9.210526 | 9 | 81 |
| 11 | 8.846761 | 10 | 100 |

**FIGURE 6.18**   First several rows of the spreadsheet for the Griffey example.

| | A | B | C | D | E | F | G |
|---|---|---|---|---|---|---|---|
| 1 | SUMMARY OUTPUT | | | | | | |
| 2 | | | | | | | |
| 3 | *Regression Statistics* | | | | | | |
| 4 | Multiple R | 0.739325876 | | | | | |
| 5 | R Square | 0.546602752 | | | | | |
| 6 | Adjusted R Sq | 0.49622528 | | | | | |
| 7 | Standard Error | 1.391398725 | | | | | |
| 8 | Observations | 21 | | | | | |
| 9 | | | | | | | |
| 10 | ANOVA | | | | | | |
| 11 | | *df* | *SS* | *MS* | *F* | *Significance F* | |
| 12 | Regression | 2 | 42.01154381 | 21.0057719 | 10.85014252 | 0.000809673 | |
| 13 | Residual | 18 | 34.84782743 | 1.935990413 | | | |
| 14 | Total | 20 | 76.85937124 | | | | |
| 15 | | | | | | | |
| 16 | | *Coefficients* | *Standard Error* | *t Stat* | *P-value* | *Lower 95%* | *Upper 95%* |
| 17 | Intercept | 2.693482545 | 1.005091305 | 2.67983867 | 0.015291999 | 0.581864071 | 4.80510102 |
| 18 | T | 0.957362712 | 0.210439016 | 4.549359375 | 0.000248447 | 0.515246745 | 1.39947868 |
| 19 | T-squared | -0.043263378 | 0.009289902 | -4.657032602 | 0.00019608 | -0.062780739 | -0.023746018 |

**FIGURE 6.19**    Output from the quadratic regression in the Griffey example.

squared. For instance, in the Griffey example, the $X$ range is B1:C22. The output for the regression procedure in this example is given in Figure 6.19. Note that it has the same general form as in the case of a single predictor, but now there is a third row in the table containing the coefficients. Based on these results, we see that the regression model is

$$Y = 2.69 + 0.957\ T - 0.0433\ T^2$$

with an $R^2$ value of 54.7%. The margins of error for the coefficients of $T$ and $T^2$ are about 0.42 and 0.019, respectively, and both coefficients are statistically significantly different from 0.

A convenient way of investigating the possibility of using a polynomial regression model or transforming either the response variable or the predictor variable is to use the "trendline" facility on a scatterplot. Consider the example of modeling points per game in terms of average time on ice for NHL forwards, discussed in Section 6.7. With the plot of points per game versus average TOI highlighted, the "Chart Tools" option is available in the toolbar. An option under Chart Tools is "Layout," and "Trendline" is a button under "Layout" (see Figure 6.20).

Clicking the trendline button leads to a menu with the entry "More trendline options"; clicking that button yields the dialog box given in Figure 6.21. Note that there are several options for the trendline, corresponding to different transformations of $X$ and/or $Y$. "Linear" corresponds to a standard linear regression; highlighting the "Linear" button and closing the dialog box adds the least-squares regression line to the chart. Checking the "Display Equation on chart" box adds the equation of the line as well, and this is a simple way to find the value of the regression line.

The "Exponential" option corresponds to a log transformation of $Y$, leaving $X$ untransformed; that is, the regression equation is of the form

$$\ln(Y) = a + b\ X;$$

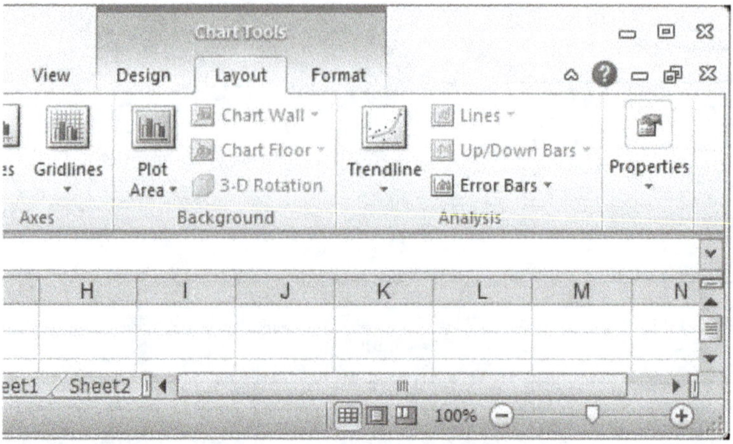

**FIGURE 6.20**  The trendline button.

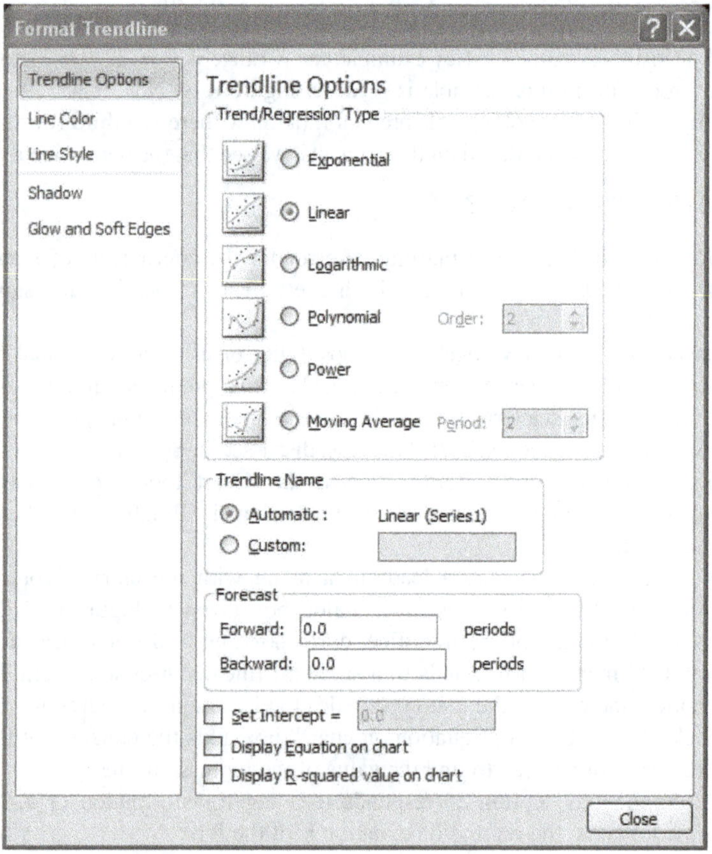

**FIGURE 6.21**  The trendline dialog box.

when this equation is displayed on the chart, and when the equation is given, it is in terms of the untransformed $Y$ so that it is of the form

$$Y = c \, e^{bX},$$

where $c = e^a$. The "Logarithmic" option corresponds to a log transformation of $X$ and no transformation of $Y$ so that the equation is of the form

$$Y = a + b \ln(X).$$

The "Polynomial" option corresponds to a polynomial regression model; the order of the polynomial can be specified. For example, choosing the order to be 2 yields the quadratic regression model

$$Y = a + b_1 X + b_2 X^2.$$

The "Power" option corresponds to a log transformation for both $Y$ and $X$, so that the model is of the form

$$\ln(Y) = a + b \ln(X).$$

Again, on the plot, the relationship is expressed in terms of the untransformed $Y$ so that the relationship has the form

$$Y = c \, X^b$$

where $c = e^a$.

Using the trendline facility therefore allows us to "try out" a large number of different models. Of course, once a model is considered to be worthy of further investigation, the Regression procedure should be used to obtain parameter estimates, margins of error, $R^2$, and so on.

# 6.13 SUGGESTIONS FOR FURTHER READING

Simple linear regression models are discussed by Agresti and Finlay (2009, Chapter 9), McClave and Sincich (2006, Chapter 9), and Snedecor and Cochran (1980, Chapter 9).

Good nontechnical discussions of regression to the mean and the regression fallacy are given by Mlodinow (2008, Chapters 1 and 8) and Gilovich (1991, Chapter 2). Regression to the mean is an important factor in many analyses of sports data; see, for example, the work of Silver (2006b) and Winston (2009, Chapter 12). The football coaches example in Section 6.7 is based on the work of Joyner (2008, Chapter 7); the data analyzed here are an updated version of the data analyzed by Joyner.

Agresti and Finlay (2009) discuss polynomial regression models in their Section 14.5 and transformations in their Section 14.6. Snedecor and Cochran (1980, Chapter 19) describe the use of regression models in describing nonlinear relationships using both transformations and polynomial regression models.

Mosteller and Tukey's (1977) book contains a comprehensive treatment of data analysis using regression models with a particularly good discussion of the usefulness of transformations, which they call "re-expression" of the variables.

Regression models in which the error terms may be correlated are discussed in Chapter 16 of the work of Agresti and Finlay (2009). The method of estimation is known as "generalized least squares"; see Freedman's (2009) Chapter 5.

# Regression Models with Several Predictor Variables

# 7

## 7.1 INTRODUCTION

The simple linear regression model relates a response variable $Y$ to a predictor variable $X$. However, often there are several variables $X_1, \ldots, X_p$ that may affect $Y$; hence, we need an extension of the simple linear regression that allows for multiple predictors.

The simple linear regression model states that the average value of $Y$ depends on the corresponding value of $X$ through the linear relationship $a + bX$, where $a$ and $b$ are constants; we write this as

$$Y = a + bX + \varepsilon,$$

where $\varepsilon$ represents "random error" that is, on average, 0.

The *multiple regression model* states that the average value of $Y$ is a linear function of $X_1, \ldots, X_p$, given by $a + b_1 X_1 + \cdots + b_p X_p$, where $a, b_1, \ldots, b_p$ are constants. As with the simple linear regression model, we may write the relationship between $Y$ and $X_1, \ldots, X_p$ as

$$Y = a + b_1 X_1 + \cdots + b_p X_p + \varepsilon$$

$$Y = a + b_1 X_1 + \cdots + b_p X_p + \varepsilon$$

where $\varepsilon$ represents random error or as

$$E(Y) = a + b_1 X_1 + \cdots + b_p X_p,$$

as discussed in Chapter 6.

# 7.2 MULTIPLE REGRESSION ANALYSIS

Many aspects of multiple regression analysis are straightforward extensions of the methods used to analyze the simple linear regression model. For instance, the function of the form $a + b_1 X_1 + \cdots + b_p X_p$ that best fits the data is chosen by least squares, that is, by minimizing the sum of squared errors, calculated by taking the difference between each $Y$ value and the corresponding value of the function $a + b_1 X_1 + \cdots + b_p X_p$, squaring this value, and adding across the set of data points.

An overall measure of how well the regression line

$$\hat{Y} = a + b_1 X_1 + \cdots + b_p X_p$$

fits the data is given by the $R^2$ *value* for the regression, which measures the relative contribution of the component $a + b_1 X_1 + \cdots + b_p X_p$ to the values of $Y$. As in the simple linear regression model, the values of $R^2$ fall between 0% and 100%, with a value of 0% indicating that $a + b_1 X_1 + \cdots + b_p X_p$ is of no help in describing $Y$ and a value of 100% indicating that $Y$ is completely determined by $a + b_1 X_1 + \cdots + b_p X_p$. Further discussion of the meaning of $R^2$ in a multiple regression analysis is given in Section 7.6.

As in a simple linear regression analysis, each coefficient $b_1, \ldots, b_p$ has an associated margin of error that can be used to calculate a range of reasonable values for the "true" regression coefficient and to determine statistical significance. If a coefficient is smaller (in magnitude) than its margin of error, then it is not statistically significant; this suggests that the corresponding $X$ variable is not needed in the regression equation describing $Y$. Techniques for choosing the $X$ variables to include in a multiple regression model are discussed in Section 7.15.

There are several ways in which a multiple regression analysis differs from a simple linear regression. One is that interpreting the regression coefficients raises some important issues. We can interpret the regression function by considering the effect of each predictor on the response, or we can think of the relationship between the entire set of predictors and the response variable; these are discussed in Sections 7.3 and 7.6, respectively. Also, the very fact that there are several predictor variables leads to new issues that do not naturally arise with only a single predictor; many of these are discussed in the remainder of this chapter.

# 7.3 INTERPRETING THE COEFFICIENTS IN A MULTIPLE REGRESSION MODEL

One way to interpret the multiple regression model is to consider the relationship between $Y$ and one of the predictor variables, holding the other predictor variables constant. For instance, consider the relationship between $Y$ and $X_1$, holding $X_2, \ldots, X_p$ constant. According to the multiple regression model,

$$Y = a + b_1 X_1 + (b_2 X_2 + \cdots + b_p X_p) + \varepsilon$$

so that the relationship between $Y$ and $X_1$ for fixed $X_2,\ldots, X_p$ is linear with slope $b_1$ and intercept $a + b_2X_2 + \cdots + b_pX_p$. Therefore, the parameter $b_1$ may be interpreted as the change in the average value of $Y$ corresponding to a 1-unit increase in $X_1$, holding $X_2,\ldots, X_p$ constant.

The interpretation of $b_1$ in the multiple regression model can be contrasted with the interpretation of $b$, the coefficient of $X_1$ in the simple linear regression model

$$Y = a + bX_1 + \varepsilon$$

relating $Y$ and $X_1$, that is, in the model with only $X_1$ as a predictor variable. The parameter $b_1$ describes the relationship between $Y$ and $X_1$ for fixed values of $X_2,\ldots, X_p$, and $b$ describes the relationship between $Y$ and $X_1$, ignoring $X_2,\ldots, X_p$. Therefore, the interpretation of the regression coefficient of a predictor variable $X_1$ in a multiple regression model depends on which other variables are in the model. This fact is crucial to properly understand the results of a multiple regression analysis.

For example, consider MLB (Major League Baseball) pitchers in the 2009 season with at least 40 innings pitched (Dataset 7.1) and consider a linear regression model with runs allowed per 9 innings ($R$) as the response variable and strikeouts per 9 innings ($K$) as the predictor variable. Using data from the 2009 season ($n = 393$), a regression analysis yields

$$\hat{R} = 6.41 - 0.260\ K.$$

Therefore, an increase of 1 strikeout per 9 innings corresponds to a decrease of 0.260 runs allowed per 9 innings.

Suppose that pitchers with more strikeouts also have more walks. Then, the effect of increasing strikeouts on runs allowed in our model might be partially offset by an increase in walks. Let $W$ denote walks per 9 innings. Then, using the data described previously, $K$ and $W$ have a correlation of 0.148, so, in fact, pitchers with more strikeouts do tend to yield more walks, on average. Suppose that we are interested in the relationship between runs allowed and strikeouts, holding walks allowed constant. One way to investigate this would be to consider a set of pitchers, all with the same number of walks allowed per 9 innings, and look at the relationship between runs allowed per 9 innings and strikeouts per 9 innings for those pitchers.

Although such a procedure could be done in principle, there are practical difficulties. In particular, relatively few pitchers have the same number of walks allowed or even have approximately the same number of walks allowed. However, the multiple regression model gives us a way to achieve the same goal using statistical methods. A second regression analysis, with response variable $R$ and predictors $K$ and $W$ leads to the multiple regression equation

$$\hat{R} = 5.24 - 0.295\ K + 0.401\ W.$$

The margins of error of the coefficients of $K$ and $W$ are about 0.06 and 0.1, respectively, so that both of the coefficients are clearly statistically significant. Therefore, an increase of 1 strikeout per 9 innings corresponds to a decrease of 0.295 runs allowed per 9 innings, *holding walks allowed per 9 innings constant*. Thus, the effect of strikeout rate on runs allowed is slightly greater when walk rate is held constant than when walk rate is ignored. Of course, it would be preferable to actually do an experiment in which $W$ is held constant. However, when this is not possible, as in the present example, the statistical approach is a useful alternative.

Note that although we are "holding $W$ constant," we never have to give the value at which it is being held. In fact, according to the multiple regression model, the effect of increasing $K$ by 1 unit is the same for all values of $W$. This is an important assumption: When it is valid, we say that there is *no interaction* between the predictors. Interaction occurs whenever the effect of one predictor depends on the values of the other predictors. The presence of interaction complicates the relationships between the variables, but it sometimes leads to more interesting conclusions. Models that allow for interaction are discussed in Sections 7.8–7.12.

Therefore, we have two different measures of the relationship between strike-outs per 9 innings and runs allowed per 9 innings. In the model ignoring walks per 9 innings (as well as all other variables not in the model), the coefficient of $K$ is −0.260; in the model that includes $W$, the coefficient of $K$ is −0.295, representing the relationship between $K$ and $R$, holding $W$ constant. Therefore, these two coefficients of $K$ have different interpretations. The difference in interpretation is important because pitchers who have more strikeouts generally have more walks.

The numerical values of the two coefficients of $K$ are very close, −0.260 and −0.295. This occurs whenever the linear relationships between the predictor variables are weak. In particular, if a given set of data $X_1$ and $X_2$ are uncorrelated, that is, their correlation is 0, then the estimate of $b_1$ in the model

$$Y = a + b_1 X_1 + b_2 X_2 + \varepsilon$$

will be the same as the estimate of $b$ in the model

$$Y = a + bX + \varepsilon.$$

However, if there are strong linear relationships between the predictor variables, the results from different regression models might be quite different. This is not unexpected because, if the predictor variables are strongly related, the implications of changing one variable while holding others constant will be important.

For instance, consider a regression model relating points scored per game ($P$) to time of possession per game ($T$) for NFL (National Football League) teams. Using data from the 2009–2011 seasons (Dataset 7.2), a linear regression analysis yields

$$\hat{P} = -17.9 + 1.32 \ T;$$

the margin of error of the coefficient of $T$ is about 0.51, so the coefficient is statistically significant. Therefore, an additional minute of possession corresponds to an additional 1.32 points, on average.

Now, consider a second predictor variable, yards gained per game ($D$). Time of possession and yards per game are fairly highly correlated, with a correlation of 0.55, indicating that teams with a higher time of possession tend to gain more yardage.

A regression analysis with points per game as the response and time of possession and yards per game as predictors yields

$$\hat{P} = -11.9 - 0.014\ T + 0.101\ D.$$

According to this analysis, more time of possession corresponds to *fewer* points per game. However, in understanding this result, it is important to keep in mind that this effect applies for fixed values of yards per game. That is, this result says that, for a fixed number of yards gained, teams score more if their yards gained occur with a lower time of possession. Although it is difficult to be certain about what is causing this result, it might be that gaining yardage through a small number of big plays tends to lead to more points than gaining yardage through a long drive.

This example also illustrates a second issue that often arises with highly correlated predictors: Although the coefficient of $T$ is statistically significant in the model with just $T$ as a predictor, it is not significant in the model with $T$ and $D$ as predictors. That is, for fixed values of the variable $D$, the variables $P$ and $T$ are not strongly related. Stated another way, time of possession does not provide any additional information about points scored that is not already reflected in yards per game. This conclusion can also be expressed using partial correlation, as discussed in Section 5.6. The partial correlation of $P$ and $T$, controlling for $D$, is only about 0.007. On the other hand, the partial correlation of $P$ and $D$, controlling for $T$, is 0.80, indicating that points scored and yards gained are linearly related, even after controlling for time of possession.

Therefore, the issue of how to treat predictor variables whose coefficients are not statistically significant is a complicated one. On the one hand, if a coefficient is not statistically significant, then there is little or no evidence that the coefficient is not 0. Therefore, it might make sense to remove that variable from the model. However, doing so changes the interpretation of the other coefficients. It follows that the goals of the analysis play an important role in the decision.

If we are simply trying to find an equation to explain the response variable $Y$ in terms of the predictors $X_1, X_2, \ldots, X_p$, it is often reasonable to remove from the model any variable whose coefficient is not statistically significant. The issue of choosing predictor variables to include in a model in this context is the subject of Section 7.15.

On the other hand, suppose our main goal is to study the relationship between $Y$ and one of the predictors, say $X_1$. Then, in describing this relationship, we might consider it important to hold certain variables constant; in this case, we may want to include these variables in the model even if their coefficients are not statistically significant.

# 7.4 MODELING STRIKEOUT RATE IN TERMS OF PITCH VELOCITY AND MOVEMENT

The PITCHf/x system, which has been installed in all major league ballparks, provides information on the speed and location of all pitches thrown. In this section, we use these data to model a pitcher's strikeout rate in terms of the average velocity of his fastball and curveball and the average movement of his curveball. Note that although data on many other types of pitches are available, identification of pitches beyond fastballs and curveballs tends to be unreliable.

This model is based on data from the 2010–2012 seasons; in a given season, all pitchers with at least 60 innings pitched are included. The variables analyzed are strikeout rate in strikeouts per 9 innings pitched ($K$), average fastball velocity in miles per hour ($F$), average curveball velocity in miles per hour ($C$), and movement of the curveball in inches ($M$). The values of $M$ tend to fall in the range 5–10, with a median value of about 8.

Data were taken from the FanGraphs.com website. Unfortunately, the data available contain many missing values. Any pitcher with a missing value for one of the variables was deleted from the dataset; this reduced the dataset by about 30% to 556 observations. Knuckleballers Tim Wakefield and R. A. Dickey were also removed from the dataset, as was Jamie Moyer, whose fastball velocity is far below the typical value for a major league pitcher.

As expected, $F$, $C$, and $M$ are each correlated with $K$. The correlation is the largest with $F$, at 0.48, followed by $C$ at 0.27 and $M$ at 0.12. The variables $F$ and $M$ are nearly uncorrelated, with a correlation of −0.002. That is, there is no (linear) relationship between a pitcher's fastball velocity and the movement on his curveball; a plot of the data (not shown) does not contain evidence of any type of relationship. Curveball velocity is positively correlated with fastball velocity ($r = 0.60$) but negatively correlated with curveball movement ($r = −0.30$); that is, pitchers throwing slower curveballs tend to have more movement on the curveball.

The data described were analyzed using a regression model with response variable $K$ and predictor variables $F$, $C$, and $M$. The resulting regression equation is

$$\hat{K} = -25.4 + 0.342\ F + 0.0106\ C + 0.0788\ M;$$

the $R^2$ value for the regression is 24.3%.

As expected, a higher fastball velocity corresponds to a higher strikeout rate for the pitcher. The same is true for curveball velocity, but the coefficient of $C$ is much lower than that of $F$. Furthermore, the margin of error of the coefficient of $C$ is about 0.05, so the coefficient is not statistically significantly different from 0. The coefficient of $M$ is small in magnitude given the range of typical values for $M$, but its margin of error is also about 0.05, so it is statistically significant. Therefore, although $C$ is more highly correlated with $K$ than $M$ is, it does not have a significant relationship with $K$ when fastball velocity is held constant, while $M$ does. This is most likely because $F$ and $M$ are unrelated, that is, they represent different aspects of a pitcher's skills, and $F$ and $C$ are highly correlated.

The regression model for $K$ in terms of $F$ and $M$ alone is

$$\hat{K} = -25.4 + 0.351\ F + 0.0748\ M,$$

with $R^2 = 24.3\%$, unchanged from the previous model. According to this model, increasing fastball velocity by 5 mph, holding curveball movement constant, corresponds to an increase of about 1.8 strikeouts per 9 innings pitched. Increasing curveball movement by 2 inches (roughly the difference between a median value of $M$ and the upper quartile), holding fastball velocity constant, corresponds to an increase of about 0.15 strikeouts per 9 innings pitched. Furthermore, it is the movement of a curveball, not its velocity, that is more relevant for predicting strikeout rate, a not-unexpected conclusion.

# 7.5 ANOTHER LOOK AT THE RELATIONSHIP BETWEEN PASSING AND SCORING IN THE ENGLISH PREMIER LEAGUE

In Section 6.9, we looked at the relationship between goals scored per game ($G$) and pass success rate ($P$) for teams in the English Premier League (EPL) for the 2009–2010 through 2012–2013 seasons. A quadratic regression function was used, with the estimated function given by

$$\hat{G} = 12.0 - 0.329\,P + 0.00247\,P^2.$$

Shots are also closely related to pass success rate. Let $S$ denote shots per game. Then, the correlation of $S$ and $P$ is about 0.66. Of course, shots are also closely related to goals scored. The relationship between $G$ and $S$ is an approximately linear one, with a correlation of about 0.79. Consider the theory that the reason that better passing corresponds to more goals scored is that better passing corresponds to more shots being taken, which in turn corresponds to more goals being scored.

To investigate this theory, we can fit the model

$$G = b_0 + b_1 P + b_2 P^2 + b_3 S + \varepsilon.$$

In this model, the coefficients $b_1$ and $b_2$ describe the relationship between goals scored per game and pass success rate for fixed values of shots taken. Therefore, if the theory described is correct, $b_1$ and $b_2$ will not be significantly different from 0.

Using the data described in Section 6.9, the estimated regression equation is

$$\hat{G} = 7.19 - 0.209\,P + 0.00150\,P^2 + 0.0958\,S.$$

The margins of error of the coefficients of $P$ and $P^2$ are roughly 0.20 and 0.0013, respectively. It follows that both estimates are statistically significantly different from 0. The margin of error of the coefficient of $S$ is roughly 0.028. It follows that pass success rate and goals scored per game are related, even after controlling for shots taken per game. That is, although the number of shots taken is related to pass success rate, for a fixed number of shots per game, a higher pass success rate still corresponds to more goals scored. The underlying reason for this result is not clear. It could be that a higher pass success rate is indicative of a higher team skill level, which results in higher-quality shots or more accurate shots.

We may view the estimated regression function as representing a quadratic relationship between goals scored and pass success, for fixed values of shots, where the constant in the quadratic depends on the shots taken but the coefficients of $P$ and $P^2$ do not. For instance, if a team averages 15 shots per game, the equation relating pass success and goals scored per game is

$$\hat{G} = 7.19 - 0.209\, P + 0.00150\, P^2 + 0.0958\, (15) = 8.63 - 0.209\, P + 0.00150\, P^2.$$

The results from the analysis in this section can be compared to the results of the analysis in Section 6.9. According to the regression equation in the present section, the estimated difference in scoring corresponding to changing the pass success rate from 65% to 70% is only about 0.03 goals per game, and the estimated difference corresponding to changing the pass success rate from 85% to 90% is about 0.3 goals per game. These can be compared to the values 0.02 and 0.5, respectively, obtained in Section 6.9. Therefore, holding shots constant has only a minor effect on the results for poorly passing teams, but it reduces the estimated change in scoring for skilled passing teams by about 0.2 goals per game, an important difference in soccer.

In interpreting these results, it is important to keep in mind that the data were collected at the season level, not at the game level. Therefore, the values of goals per game, shots per game, and pass success rate refer to a team's skill level rather than to their performance in a specific game.

# 7.6 MULTIPLE CORRELATION AND REGRESSION

In Chapter 5, several different definitions of correlation were presented. The goal of these measures is to summarize the strength of the linear relationship between two variables. In this section, the results of a multiple regression analysis are interpreted as the solution to a correlation problem.

Let $Y$ be a response variable of interest and let $X_1, \ldots, X_p$ denote variables thought to be related to $Y$. Suppose we want to measure the strength of the linear relationship between $Y$ and the set $X_1, \ldots, X_p$. One answer to this question is provided by the set of correlation coefficients of $Y$ and each of $X_1, \ldots, X_p$. However, those values do not address the issue of how $Y$ relates to the entire set $X_1, \ldots, X_p$. Therefore, we consider an alternative approach.

For a given set of constants $g_1, \ldots, g_p$, consider the function

$$g_1 X_1 + \cdots + g_p X_p.$$

One way to measure the strength of the linear association between $Y$ and $X_1, \ldots, X_p$ is to find the constants $g_1, \ldots, g_p$ that maximize the correlation of $g_1 X_1 + \cdots + g_p X_p$ and $Y$; the value of this correlation is called the *multiple correlation* of $Y$ and $X_1, \ldots, X_p$, and it is generally denoted by $R$.

As the notation suggests, $R$ is simply the square root of the $R^2$ value (expressed as a proportion rather than as a percentage) for the regression of $Y$ on $X_1, \ldots, X_p$.

Furthermore, the linear function of $X_1,\ldots,X_p$ maximizing the correlation with $Y$ is given by the regression coefficients for the regression of $Y$ on $X_1,\ldots,X_p$, $b_1X_1 + \cdots + b_pX_p$. However, because correlation is unaffected by rescaling the variables, $b_1X_1 + \cdots + b_pX_p$ gives the relative contribution of the variables, rather than the unique linear function. That is, $c(b_1X_1 + \cdots + b_pX_p)$ for any constant $c$ has the same correlation with $Y$ as does $b_1X_1 + \cdots + b_pX_p$.

Consider the following example: Let the response variable be the runs scored by a given team in a particular season, denoted by $Y$, and let the predictors be other measures of offensive performance: hits ($H$), doubles ($D$), triples ($T$), home runs ($HR$), walks ($W$), and the number of times a player has grounded into a double play ($DP$). Each of these variables is measured on a per plate appearance basis for the team. Using data from 5 seasons, 2007–2011 (Dataset 7.3), the estimated regression model is

$$\hat{Y} = -894 + 4900\ H + 1518\ D + 3947\ T + 5645\ HR + 3616\ W - 2636\ DP.$$

The right-hand side of this equation is the linear function of $H, D, T, HR, W$, and $DP$ that has maximum correlation with $Y$; this correlation is 0.960.

The predictor variables included in this analysis are similar to those used in calculating WOBA (weighted on-base average; also often denoted wOBA), designed to measure offensive performance; hence, it is useful to compare the functions. WOBA is given by

$$0.90S + 1.24D + 1.56T + 1.95\ HR + 0.72NIW + 0.75\ HBP + 0.92\ E$$

where $S$ denotes singles, $NIW$ denotes nonintentional walks, $HBP$ denotes hit by pitch, and $E$ denotes reached base on error (see Chapter 1 of the 2007 work by Tango, Lichtman, and Dolphin).

Obviously, the coefficients in $\hat{Y}$ are much larger than those in WOBA, making direct comparison pointless. However, as noted previously, we may rescale the coefficients in the regression function without altering its correlation-maximizing property. Using the fact that $H = S + D + T + HR$, ignoring the constant $-894$, which does not affect correlation, and rescaling the expression for $\hat{Y}$ so that the coefficient of $S$ is 0.90, the linear function of $S, D, T, HR, W$, and $DP$ that maximizes the correlation with runs scored is given by

$$0.90S + 1.18D + 1.63T + 1.94HR + 0.66W - 0.48DP,$$

which is in close agreement with the expression for WOBA.

# 7.7 MEASURING THE OFFENSIVE CONTRIBUTION OF PLAYERS IN LA LIGA

Soccer is a sport in which, although there is quite a bit of offensive play, there is relatively little scoring. Therefore, traditional methods of evaluating offensive play, based on goals and assists, can be misleading. In this section, we consider the problem of measuring the offensive contribution of soccer players in La Liga.

We use the same basic approach used in the previous section to measure the offensive contribution of baseball players. The first step is to find an equation relating various offensive statistics to team scoring. We can then use this equation at the player level to determine a player's contribution to his team's scoring. Because the results may be specific to a particular league, here we analyze data from La Liga only.

The offensive statistics available are crosses ($C$), through balls ($T$), long balls ($L$), passes ($P$), shots ($S$), and dribbles ($D$); these data are taken from WhoScored.com. Using team data from the 2009–2010 through 2012–2013 seasons, a regression model was fit with goals scored $G$ as the response variable and $C$, $T$, $L$, $P$, $S$, and $D$ as the predictors. The resulting regression equation is

$$\hat{G} = -53.9 - 0.0108\ C - 0.00455\ T + 0.00384\ L + 0.00241\ P + 0.103\ S + 0.0575\ D;$$

the value of $R^2$ for the regression is 78.4%. The coefficients of $C$, $T$, and $L$ are not statistically significantly different from 0. Note that passes includes crosses, through balls, and long balls, so the coefficient of $C$, for example, represents the contribution of crosses to scoring in addition to that of a standard pass. Therefore, the fact that the coefficients of $C$, $T$, and $L$ are not significant means that, although the number of successful passes is important in predicting goals scored, the type of pass is not important.

The regression equation with $G$ as the response variable and $P$, $S$, and $D$ as predictors is

$$\hat{G} = -50.8 + 0.00243\ P + 0.0969\ S + 0.0558\ D;$$

for this model, $R^2 = 78.1\%$, and all coefficients are significantly different from 0. Therefore, we can use this equation to measure the contribution of goals scored for each player. Note that, at the player level, we do not expect this equation to predict goals scored because some players are expected to pass more than score and others are expected to score more than pass. Instead, the equation gives us a way to measure a player's contribution to his team's goals scored based on his values of $P$, $S$, and $D$.

Suppose that a particular player makes $P_1$ passes, takes $S_1$ shots, and has $D_1$ dribbles for some specific respective values $P_1$, $S_1$, $D_1$; then, his contribution to the team's goals scored is $0.00243\ P_1 + 0.0969\ S_1 + 0.0558\ D_1$. However, if we add the contributions of all of a team's players, the total will exceed the team's actual goals scored by about 50.8 goals, the value of the constant in the regression equation. To obtain a player's offensive contribution to the team, measured in goals, we can use

$$OC = -5.08 + 0.00243P + 0.0969S + 0.0558D;$$

this is simply the regression equation with the constant divided by 10, reflecting the fact that there are 10 field players at any one time. Note that the value of the constant is irrelevant for comparing players.

For example, in 2012–2013, Lionel Messi made 1760 passes, took 163 shots, and made 122 dribbles. Therefore, his value of $OC$ is

$$-5.08 + 0.00243(1760) + 0.0969(163) + 0.0558(122) = 21.8.$$

**TABLE 7.1**    Top twenty players in La Liga in *OC*

| RANK | | OC |
|------|--------------------|------|
| 1 | Cristiano Ronaldo | 23.6 |
| 2 | Lionel Messi | 21.8 |
| 3 | Álvaro Negredo | 13.2 |
| 4 | Falcao | 10.5 |
| 5 | Piti | 10.5 |
| 6 | Carlos Vela | 10.4 |
| 7 | Jesús Navas | 9.4 |
| 8 | Iago Aspas | 9.2 |
| 9 | Pizzi | 9.1 |
| 10 | Antoine Griezmann | 7.8 |
| 11 | Andrés Iniesta | 7.5 |
| 12 | Roberto Soldado | 7.5 |
| 13 | Roberto Trashorras | 7.5 |
| 14 | Ander | 7.4 |
| 15 | Emiliano Armenteros | 6.9 |
| 16 | Markel Susaeta | 6.7 |
| 17 | Francesc Fàbregas | 6.6 |
| 18 | Jonas | 6.4 |
| 19 | Xavi | 6.3 |
| 20 | Éver Banega | 6.3 |

Obviously, this is much lower than the actual 46 goals that he scored. One way to view the difference is that it is because, for many of his goals, other players did much of the "work" by passing and the like. Also, this type of offensive contribution measure is likely to be more valuable for players who do not score many goals but who contribute in other ways; we do not need analytic methods to realize that Messi makes a huge contribution to his team.

Table 7.1 contains the top 20 players in La Liga in terms of *OC*. It is interesting to note that the top four goal scorers in La Liga (Messi, Cristiano Ronaldo, Falcao, and Alvaro Negredo) are also the top four in *OC*, even though the *OC* formula does not include goals scored.

# 7.8 MODELS FOR VARIABLES WITH A SYNERGISTIC OR ANTAGONISTIC EFFECT ON PERFORMANCE USING INTERACTION

Interaction occurs when the effect of one predictor on the response variable depends on the values of one or more of the other predictors. Although, in principle, interaction can be complex, fortunately there are simple models that can be used to represent relationships with interaction.

Consider modeling a response variable $Y$ in terms of predictors $X_1$ and $X_2$. The usual multiple regression model has the form

$$Y = a + b_1 X_1 + b_2 X_2 + \varepsilon;$$

such a model is said to be *additive* because the effect of $X_1$ and $X_2$ is formed by adding the effect of $X_1$ and the effect of $X_2$. As discussed in Section 7.3, one consequence of such an additive model is that the effect of increasing $X_1$ by 1 unit on the average value of $Y$ is the same for all values of $X_2$, and conversely the effect of increasing $X_2$ by 1 unit on the average value of $Y$ is the same for all values of $X_1$.

Now, suppose that the effect of increasing $X_1$ by 1 unit on the average value of $Y$ depends on the value of $X_2$. The simplest type of dependence is one in which the effect of increasing $X_1$ is linear in $X_2$, say $b_1 + b_{12} X_2$, for some parameters $b_1$, $b_{12}$; that is, suppose the coefficient of $X_1$ is of the form $b_1 + b_{12} X_2$. Then, the regression model takes the form

$$Y = a + (b_1 + b_{12} X_2) X_1 + b_2 X_2 + \varepsilon.$$

This can be rewritten as

$$Y = a + b_1 X_1 + b_2 X_2 + b_{12} X_1 X_2 + \varepsilon.$$

That is, to allow for interaction in the model, we include a third predictor variable, the product of $X_1$ and $X_2$. Under this model, the slope of the regression line relating $Y$ and $X_1$ is a linear function of $X_2$; conversely, the slope of the regression line relating $Y$ and $X_2$ is a linear function of $X_1$.

Under this model, increasing $X_1$ by 1 unit, holding $X_2$ fixed, corresponds to a change in the average value of $Y$ of $b_1 + b_{12} X_2$, and increasing $X_2$ by 1 unit, holding $X_1$ fixed, corresponds to a change in the average value of $Y$ of $b_2 + b_{12} X_1$. On the other hand, consider increasing both $X_1$ and $X_2$ by 1 unit. Then, the average value of $Y$ changes from

$$a + b_1 X_1 + b_2 X_2 + b_{12} X_1 X_2$$

to

$$a + b_1 (X_1 + 1) + b_2 (X_2 + 1) + b_{12} (X_1 + 1)(X_2 + 1),$$

a change of $b_1 + b_2 + b_{12}(X_1 + X_2 + 1)$.

Therefore, if $b_{12} > 0$, the effect of increasing both $X_1$ and $X_2$ is greater than the sum of the effects of increasing each of $X_1$ and $X_2$ separately. In this case, the effects of $X_1$ and $X_2$ are said to be *synergistic*. On the other hand, if $b_{12} < 0$, the effect of increasing both $X_1$ and $X_2$ is less than the sum of the effects of increasing each of $X_1$ and $X_2$ separately. In this case, the effects of $X_1$ and $X_2$ are said to be *antagonistic*. When $b_{12} = 0$, the model is additive, as discussed previously.

Consider the example analyzed in Section 7.3 in which the response variable $R$, runs allowed per 9 innings, is modeled as a function of $K$, strikeouts per 9 innings, and $W$, walks per 9 innings. The regression equation, using data on pitchers in the 2009 season with at least 40 innings pitched (Dataset 7.1), is

$$\hat{R} = 5.24 - 0.295\ K + 0.401\ W.$$

According to this model, pitchers who allow more walks give up more runs and pitchers who strike out more batters allow fewer runs. Furthermore, in this model, the effects of strikeouts and walks allowed are additive, so that, for example, increasing walks allowed per 9 innings by 1 corresponds to a 0.401 increase in runs allowed per 9 innings, and that value applies to any level of strikeouts per 9 innings.

To determine if there is interaction between $K$ and $W$, we can fit a model with three predictor variables, $K$, $W$, and $K*W$, the product of $K$ and $W$. Note that in specific examples, we write the interaction term with the multiplication symbol * to avoid confusion with a predictor variable denoted by two letters such as $KW$. The resulting regression equation is

$$\hat{R} = 4.73 - 0.226\ K + 0.546\ W - 0.0196\ K*W.$$

The margin of error of the estimate of the coefficient of $K*W$ is 0.053; therefore, the estimate is not statistically significant, and there is no evidence of interaction between $K$ and $W$.

When interaction is present, more care is needed in interpretation of the regression coefficients. For instance, in a model with predictors $X_1, X_2$ and $X_1 X_2$, we cannot increase $X_1$ (or $X_2$) by 1 unit while holding $X_1 X_2$ fixed. See the examples in Sections 7.9 and 7.10 for detailed discussion of the interpretation of models with interaction.

Although here we have discussed interaction in models with two predictors, the same concepts apply to regression models with many predictors. Consider a model with response variable $Y$ and predictors $X_1, X_2, \ldots, X_p$. In principle, we might have interaction between any pair of predictors. Therefore, there are potentially $p(p-1)/2$ interaction terms that could be included in the model. However, once $p$ goes beyond 2 or 3, it is impractical to consider all possible interaction terms, and it is important to only include those interaction terms whose presence makes sense in the context of the model and the data being analyzed.

# 7.9 A MODEL FOR 40-YARD DASH TIMES IN TERMS OF WEIGHT AND STRENGTH

Consider the example on the 40-yard dash times and the weights of participants in the NFL Scouting Combine, discussed in Section 6.10. In that section, models with different combinations of log transformations of the variables were compared, and the model with response variable $\ln(T)$, where $T$ denotes a player's 40-yard dash time in seconds, and predictor variable $\ln(W)$, where $W$ denotes the player's weight in pounds, was chosen. Using data from the 2008–2011 combines, the estimated regression line is

$$\widehat{\ln(T)} = -0.118 + 0.306\ln(W).$$

According to this result, a 5% increase in weight corresponds to roughly a 1.5% increase in 40-yard dash time; see Section 6.10 for the details on how to interpret the results from models with log-transformed variables.

In some cases, heavy players compensate for their weight by being stronger. Therefore, in understanding the factors affecting 40-yard dash time, it may be useful to include some measure of strength in the model. At the combine, many players participate in the bench press test, in which the number of times they can bench press 225 pounds is recorded; call this variable $B$. Note that although 1321 players recorded 40-yard dash times and weights in the 2008–2011 combines, only 1061 of these recorded a bench press score; those without a bench press score were deleted from the dataset. Although $B$ could be transformed, here we use $B$ itself as the predictor.

The estimated regression model with $\ln(T)$ as the response variable and $\ln(W)$ and $B$ as predictor variables is given by

$$\widehat{\ln(T)} = -0.334 + 0.351\ln(W) - 0.00147\ B.$$

According to this equation, increasing a player's weight by 5% while holding bench press score constant corresponds to roughly a 1.7% increase in 40-yard dash time. This is slightly higher than the result in the model ignoring bench press score, which makes sense because increasing $W$ by 5% pounds while holding $B$ constant means a heavier, but not stronger, player. As in the previous model, both estimated coefficients are statistically significant; the margin of error of the $\ln(W)$ coefficient is about 0.014, and the margin of error of the $\ln(B)$ coefficient is about 0.00040.

When the coefficients of some of the predictors are highly significant, as in this case, it is often a good idea to consider the possibility of interaction. For a model with response $\ln(T)$ and predictors $\ln(W), B$ and $B^*\ln(W)$, the estimated regression model is

$$\widehat{\ln(T)} = 0.0667 + 0.278\ \ln(W) - 0.0204\ B + 0.00343\ B^*\ln(W)\ ;$$

the margin of error of the interaction coefficient is 0.0018, so the coefficient is statistically significant. Therefore, the effect of increasing weight depends on the strength of the player, as measured by $B$.

To illustrate the effect of the interaction, consider two players, a "strong" one, with $B = 30$, and a "weak" one, with $B = 10$. For a strong player, substituting 30 for $B$ in the regression equation yields

$$\widehat{\ln(T)} = -0.545 + 0.381\ln(W).$$

It follows that, for a strong player, an increase in $W$ of 5% corresponds to an increase in the player's 40-yard dash time of about 1.9%; note that $0.381 \times 0.05 = 0.019$.

For a weak player,

$$\widehat{\ln(T)} = -0.137 + 0.312 \ln(W)$$

so an increase in $W$ of 5% corresponds to an increase in the player's 40-yard dash time of about 1.5%.

These results are somewhat counterintuitive: They say that the effect of increased weight is greater on a strong player than on a weak one. However, when interpreting interaction, it is important to keep in mind it applies in both directions. For instance, in the example, the interaction says that the effect of weight depends on strength; it also says that the effect of strength depends on weight.

Consider two players, a heavy player with $W = 350$ and a light player with $W = 200$. For the light player, substituting $\ln(W) = 5.30$ in the regression equation yields

$$\widehat{\ln(T)} = 1.542 - 0.00223 \ B.$$

Therefore, an increase in $B$ of 5 corresponds to a decrease in 40-yard dash time of about 1.1%. For the heavy player, $\ln(W) = 5.858$, leading to the equation

$$\widehat{\ln(T)} = 1.695 - 0.000257 \ B.$$

According to this equation, an increase in $B$ of 5 corresponds to little change in the 40-yard dash time, specifically, a decrease of about 0.13%.

That is, for a light player, greater strength corresponds to more speed; however, for a heavy player, strength has relatively little effect on speed. It is important to keep in the mind that relationships are not causal; it may be that heavier players naturally increase their strength to compensate for the added weight, and this keeps their speed constant, rather than increasing it.

As discussed in the previous section, interaction in a model often indicates the presence of either a synergistic or an antagonistic effect of the variables in the model. For instance, in the 40-yard dash example discussed, consider a 300-pound player who scores 20 on the bench press test. According to our model, a decrease of 10% in the player's weight, holding $B$ constant, corresponds to a decrease of 0.037 in $\ln(T)$, or about 3.7% in $T$. An increase of his bench press score of 10 corresponds to a decrease of 0.008 in $\ln(T)$. However, a decrease in weight of 10% *and* an increase of 10 in the bench press score corresponds to a decrease of 0.048 in $\ln(T)$. Because 0.048 is greater than $0.037 + 0.008$, there is a (small) synergistic effect of decreasing weight and increasing strength, at least according to this regression model. Also, as always, we must remember that the relationships implied by the regression models simply summarize the relationships between the variables in the dataset and are not causal.

# 7.10 INTERACTION IN THE MODEL FOR STRIKEOUT RATE IN TERMS OF PITCH VELOCITY AND MOVEMENT

In Section 7.4, we considered a regression model relating strikeout rate in strikeouts per 9 innings pitched ($K$) to average fastball velocity in miles per hour ($F$), average curveball velocity in miles per hour ($C$), and movement of the curveball in inches ($M$). See Section 7.4 for further details on the data used and the variables.

Recall that the regression model with response variable $K$ and predictor variables $F$, $C$, and $M$ is

$$\hat{K} = -25.4 + 0.342\ F + 0.0106\ C + 0.0788\ M.$$

The coefficient of $C$ is not statistically significant. The regression model for $K$ in terms of $F$ and $M$ alone is

$$\hat{K} = -25.4 + 0.351\ F + 0.0748\ M.$$

This is a case in which we might reasonably expect interaction to be a factor—a pitcher's success often depends on his ability to throw more than one pitch effectively. Therefore, the effect of one pitch might depend on his ability with other pitches.

As a starting point, we can consider a model that includes the interaction terms for the predictors $F$, $C$, and $M$; that is, we include the cross-product terms $F*C$, $F*M$, and $C*M$ as predictors. Because we should include the first-order term of any variable present in cross-product terms, we fit a regression model with response variable $K$ and predictors $F$, $C$, $M$, $F*C$, $F*M$, and $C*M$. The resulting regression equation is

$$\hat{K} = 147.91 - 1.549\ F - 2.289\ C + 0.00244\ M + 0.0251\ F*C$$

$$- 0.00187 F*M + 0.00326 C*M.$$

The coefficients of the interaction terms involving movement, $F*M$ and $C*M$, have margins of error 0.016 and 0.024, respectively; hence, neither coefficient is close to being statistically significant. Eliminating these terms from the model yields the regression equation

$$\hat{K} = 146.97 - 1.562\ F - 2.259\ C + 0.0836\ M + 0.0250\ F*C$$

with $R^2 = 26.3\%$. All coefficients in this equation are statistically significant; therefore, according to this equation, there is interaction between fastball velocity and curveball velocity.

Care is needed when interpreting the coefficients in an equation with interaction. For instance, the negative coefficient of $C$ does not necessarily mean that higher curve-ball velocity corresponds to a lower strikeout rate because slope in the relationship between $K$ and $C$ depends on the value of $F$.

For example, suppose that $F = 86$. Then, replacing $F$ with 86 in the previous regression equation yields

$$\hat{K} = 12.64 + 0.0836\ M - 0.109\ C$$

so that, for pitchers with a relatively slow average fastball velocity of 86 mph, higher curveball velocity corresponds to a lower strikeout rate. On the other hand, suppose that $F = 96$. Then,

$$\hat{K} = -2.98 - 0.0836\ M + 0.303\ C$$

so that, for pitchers with a relatively fast average fastball velocity of 96 mph, higher curveball velocity corresponds to a higher strikeout rate. Note that, in both cases, these results apply to a fixed level of average movement in the curveball.

That is, according to this model, a power pitcher, with a lot of velocity on his fastball, is more successful (in terms of strikeouts) if he also has a lot of velocity on his curveball. On the other hand, a pitcher with a slow fastball is more successful if his curveball is also slow.

# 7.11 USING CATEGORICAL VARIABLES, SUCH AS LEAGUE OR POSITION, AS PREDICTORS

So far, we have only considered regression models in which all the variables are quantitative. However, in some cases, it is natural to include a categorical variable as a predictor in a regression model.

Recall the example relating the response variable $R$, runs allowed per 9 innings; $K$, strikeouts per 9 innings; and $W$, walks per 9 innings. The regression equation, using data on pitchers in the 2009 season with 40 innings pitched (Dataset 7.1), is

$$\hat{R} = 5.238 - 0.295\ K + 0401\ W.$$

This analysis uses data from both the American and National Leagues (AL and NL, respectively). However, it is generally believed that the AL is more difficult for pitchers at least partly because of the presence of the designated hitter. Therefore, it may be useful to include the pitcher's league in the model.

Hence, we would like to be able to include the categorical variable "league" in the regression model. To do this, we need to convert it to a quantitative variable. This is done using an *indicator variable,* also called a *dummy variable.* Let $L$ denote a variable taking the value 1 if the pitcher is from the AL and value 0 if the pitcher is from the NL. Then, $L$ indicates the value of the categorical variable league; furthermore, it can be used as a quantitative variable in a regression model.

Consider a regression model of the form

$$R = a + b_1 K + b_2 W + b_3 L + \varepsilon.$$

Suppose that a pitcher is from the AL. Then, $L = 1$ and

$$R = a + b_1 K + b_2 W + b_3 + \varepsilon;$$

if a pitcher is from the NL, then $L = 0$ and

$$R = a + b_1 K + b_2 W + \varepsilon.$$

Therefore, according to this model, the effect of the variable $L$ in the model is to make the intercept term dependent on the player's league; the relationship between runs allowed and strikeouts and walks is the same for both leagues.

Using data from the 2009 season, the estimated regression model is

$$\hat{R} = 5.013 - 0.292\,K + 0411\,W + 0.354\,L.$$

The margin of error of the coefficient of league is 0.22, so the estimated coefficient is statistically significant; that is, league does have an effect on runs allowed, holding strikeouts and walks constant. Note that the coefficients of $K$ and $W$ are nearly unchanged; the effect of including league in the model is to replace the intercept 5.24 in the original model by a league-dependent intercept: 5.01 for NL pitchers and 5.36 for AL pitchers. Therefore, if we compare two pitchers, one in the AL and one in the NL, with the same strikeout and walk rates, the AL pitcher is expected to yield 0.35 more runs per game than the NL pitcher.

Interaction between an indicator variable and a quantitative variable has a particularly simple interpretation. Consider the model

$$R = a + b_1 K + b_2 W + b_3 L + b_4 K * L + b_5 W * L + \varepsilon.$$

Suppose a pitcher is from the AL. Then, $L = 1$ so that the regression model for such a pitcher is given by

$$R = (a + b_3) + (b_1 + b_4)K + (b_2 + b_5)W + \varepsilon;$$

for a pitcher from the NL, $L = 0$, so the regression model is

$$R = a + b_1 K + b_2 W + \varepsilon.$$

That is, interaction between a quantitative variable and an indicator variable corresponds to different slopes, as well as different intercepts, for the groups defined by the indicator variable.

In the present example, the estimated regression model is

$$\hat{R} = 4.966 - 0.233\ K + 0.307\ W + 0.342\ L - 0.120\ K*L + 0.274\ W*L.$$

The coefficient of $L$ in this model is not statistically significant; all other coefficients are significant. Because the interaction terms including $L$ are statistically significant, they should remain in the model. If an interaction term including two variables is in the model, it is standard practice to include the first-order terms in the model as well. That is, in the example, because the coefficients of the $K*L$ and $W*L$ terms are statistically significant, the model should include $K$, $W$, $L$, $K*L$, and $W*L$, without regard to the significance of the $K$, $W$, and $L$ coefficients.

To interpret the model, we can look at AL and NL pitchers separately. For an AL pitcher, the regression equation is

$$\hat{R} = 5.308 - 0.353\ K + 0.581\ W.$$

For an NL pitcher, the regression equation is

$$\hat{R} = 4.966 - 0.233\ K + 0.307\ W.$$

It follows that a pitcher pays a greater price for a walk in the AL than in the NL. Also, strikeouts are more beneficial to AL pitchers; based on our regression model, having 1 additional strikeout per 9 innings corresponds to a decrease of 0.353 runs allowed per 9 innings for AL pitchers; the value for NL pitchers is only 0.233 runs per 9 innings. In both cases, walks are held constant.

Therefore, a categorical variable taking two values is easily handled by using an indicator variable. Now, suppose that the categorical variable of interest has three or more categories. Let $Z$ denote the categorical variable and suppose that it takes one of the values $v_1, v_2, ..., v_m$; that is, $Z$ takes one of $m$ possible values. As was the case with the categorical variable league discussed in the previous example, to use $Z$ in a regression analysis we need to convert it to indicator variables. When $Z$ takes two possible values, that is, when $m = 2$, this conversion process is straightforward: We use an indicator variable that is equal to 1 when $Z = v_1$ and is equal to 0 when $Z = v_2$. When $Z$ takes $m$ possible values, $m > 2$, we need $m - 1$ indicator variables to describe the value of $Z$.

Let $Z_1$ be an indicator variable taking value 1 if $Z = v_1$ and taking value 0 otherwise. Let $Z_2$ be an indicator variable taking value 1 if $Z = v_2$ and taking value 0 otherwise. Define $Z_3, ..., Z_{m-1}$ similarly so that $Z_j$ is 1 if $Z = v_j$ and is 0 otherwise for each $j$ from 1 to $m - 1$. For instance, suppose that $Z = v_1$; then

$$Z_1 = 1,\ Z_2 = 0, ...,\ Z_{m-1} = 0.$$

If $Z = v_2$, then

$$Z_1 = 0,\ Z_2 = 1,\ Z_3 = 0, ...,\ Z_{m-1} = 0.$$

If $Z = v_{m-1}$, then

$$Z_1 = 0, Z_2 = 0, ..., Z_{m-2} = 0, Z_{m-1} = 1.$$

If $Z = v_m$, then all of the $Z_j$ are equal to 0. It is easy to check that knowing the value of the sequence $Z_1, Z_2, ..., Z_{m-1}$ is equivalent to knowing the value of $Z$ and vice versa.

We can include the categorical variable $Z$ in the regression model using the same basic approach used in the pitching example described previously in this section. Consider a regression model with response variable $Y$ and predictor $X$ and suppose we want to include $Z$ in the model. To do this, we use the model

$$Y = a + b_1 X + b_2 Z_1 + \cdots + b_m Z_{m-1} + \varepsilon.$$

As in the two-category case, this model represents a linear relationship between $Y$ and $X$ in which the intercept depends on the value of $Z$. For instance, if $Z = v_1$, then $Z_1 = 1$ and $Z_2 = \cdots = Z_{m-1} = 0$ so that

$$Y = a + b_2 + b_1 X + \varepsilon.$$

Note that the slope in the linear relationship between $Y$ and $X$ does not depend on the value of $Z$. If it is believed that the slope also depends on the value of $Z$, then we can include interaction terms of the form $X * Z_j$, for $j = 1, ..., m - 1$.

# 7.12 THE RELATIONSHIP BETWEEN REBOUNDING AND SCORING IN THE NBA

Consider a model designed to study the relationship between scoring and rebounding among NBA (National Basketball Association) players. Let $Y$ denote points scored per game for a given player in a season and let $X$ denote the player's offensive rebounds per game for that season. Using data on players with at least 70 games played in the 2010–2011 season (Dataset 7.4; $n = 185$), the estimated regression equation is

$$\hat{Y} = 10.18 + 1.417 \ X;$$

according to this equation, an additional offensive rebound per game is worth about 1.4 points per game over the course of a season. The margin of error of the slope estimate is 0.861, so the estimate is statistically significant; however, the $R^2$ value is relatively low, 5.6%.

It seems likely that the relationship between scoring and rebounding will depend on a player's position. Let $P$ denote the position of a player with $P = C$ for centers, $P = F$ for forwards and $P = G$ for guards; here, the position is the one assigned by ESPN.com. Therefore, $P$ is a categorical variable with three categories. To use $P$ in a regression model, we must first convert it to indicator variables. Let $P_C = 1$ if the player is a center and 0 otherwise and let $P_F = 1$ if the player is a forward and 0 otherwise. That is, the pair $(P_C, P_F)$ takes the value (1, 0) for centers, (0, 1) for forwards, and (0, 0) for guards.

The estimated regression model with predictors $X$, $P_C$, $P_F$ is given by

$$\hat{Y} = 10.54 + 2.977\ X - 6.59\ P_C - 2.73\ P_F.$$

This can be viewed as describing a separate regression equation for each position, with the slope the same for each position and the intercept depending on the position. That is, for centers, the relationship is given by

$$\hat{Y} = 3.95 + 2.977\ X;$$

for forwards, the relationship is

$$\hat{Y} = 7.81 + 2.977\ X$$

and for guards, the relationship is

$$\hat{Y} = 10.54 + 2.977\ X.$$

Therefore, under this model, the effect of position is a fixed points-per-game adjustment, not dependent on the number of rebounds.

Judging statistical significance is a little more complicated when several predictor variables jointly represent a single categorical variable. Although we have a margin of error for each coefficient estimate, these apply to the indicator variables themselves, not to the underlying categorical variables. Although tests of statistical significance are available in this setting, we do not discuss such tests here. Instead, we use an informal method based on looking at the margins of error for all the indicator variables representing the categorical variable. In the present example, the margin of error of the coefficient of $P_C$ is 2.72; the margin of error of the coefficient of $P_F$ is 1.91. Therefore, both estimated coefficients are statistically significant, so it is safe to say that the categorical variable "position" should be included in the model. Also, note that the $R^2$ value for the model including position is 16.4%, a substantial increase from the value (5.6%) for the model with only offensive rebounds per game as a predictor.

Given the significance of the coefficients of $P_C$ and $P_F$, it is worth considering possible interaction between position and offensive rebounds. The estimated regression model including the interaction terms $X * P_C$ and $X * P_F$ is given by

$$\hat{Y} = 7.85 + 7.756\ X - 5.99\ P_C + 1.10\ P_F - 3.790\ X * P_C - 5.538\ X * P_F.$$

This gives a separate linear relationship between points scored and offensive rebounds for each position. For centers, the relationship is

$$\hat{Y} = 1.86 + 3.966 \ X;$$

for forwards, the relationship is

$$\hat{Y} = 8.95 + 2.218 \ X$$

and for guards, the relationship is

$$\hat{Y} = 7.85 + 7.756 \ X.$$

Therefore, for centers, the effect of rebounding on scoring is particularly important: According to the regression equation, a center with 1 offensive rebound per game scores 5.83 points per game on average, while a center with 3 offensive rebounds per game (a reasonable value for a top rebounder) scores 13.76 points per game. Although the coefficient of $X$ is much larger for guards, it is important to keep in mind that the best-rebounding guards have values of $X$ around 1.5. Also, note that the regression equation should not be interpreted to mean that each offensive rebound leads to nearly 8 points for a guard—what it means is that guards who have many offensive rebounds also tend to be high scorers. This might be because of the type of game these players play or their physical characteristics.

To address the issue of the statistical significance of the interaction between position and offensive rebounds, we can look at the margins of error of the coefficient estimates. The coefficient of $X * P_C$ has a margin of error of 4.39, and the coefficient of $X * P_F$ has a margin of error of 3.82. Therefore, the coefficient of $X * P_F$ is statistically significant, but the coefficient of $X * P_C$ is not (although it is close). The $R^2$ value for the regression is 20.5%. Given the significance of one coefficient, the near significance of the other, and the increase in $R^2$, along with the fact that it makes sense that the relationship between offensive rebounds and scoring should be different for different positions, it is appropriate to include the interaction terms in the model.

Finally, it is worth noting that the conclusions of the analysis do not depend on which values of the categorical variables are used to form the indicator variables for the regression. For instance, in the example, let $P_G$ take the value 1 if the player is a guard and 0 otherwise. Then, the estimated regression model using predictors $X, P_C, P_G$ is given by

$$\hat{Y} = 7.82 + 2.977 \ X - 3.86 \ P_C + 2.73 \ P_G.$$

Recall that the estimated regression model with predictors $X, P_C, P_F$ is

$$\hat{Y} = 10.54 + 2.977 \ X - 6.59 \ P_C - 2.73 \ P_F.$$

Although these equations look different, in fact they imply the same relationships among the variables.

For instance, consider the relationship between points scored and offensive rebounds for guards. Using the model with predictors $X$, $P_C$, $P_F$, with $P_C = P_F = 0$, the equation is

$$\hat{Y} = 10.54 + 2.977\ X.$$

Using the model with predictors $X$, $P_C$, $P_G$, with $P_C = 0$ and $P_G = 1$, we have

$$\hat{Y} = 7.82 + 2.977\ X - 3.86\ (0) + 2.73(1)$$
$$= 7.82 + 2.73 + 2.977\ X = 10.55 + 2.977\ X.$$

That is, the estimated relationship between points scored and offensive rebounds is the same for the two cases. This is true in general: Although the regression equations may appear to be different, the conclusions of the analysis do not depend on how the categorical variable is represented in the regression equation.

## 7.13 IDENTIFYING THE FACTORS THAT HAVE THE GREATEST EFFECT ON PERFORMANCE: THE RELATIVE IMPORTANCE OF PREDICTORS

When modeling a response variable $Y$ in terms of predictor variables $X_1,\ldots, X_p$, it is natural to consider the relative importance of different predictors. For instance, if we are modeling the performance of a player (or team) in terms of several characteristics of the player, it is useful to know which characteristics are most important in player performance. Unfortunately, this simple question does not have a simple answer. In this section, several methods of assessing the relative importance of predictors are presented, differing in the way that "importance" is interpreted.

Consider the following example: Let the response variable be the points scored per game ($P$) of an NBA team in a given season and take as the predictor variables the field goal shooting percentage ($S$) and the free-throw percentage $F$ of the team in the given season. Using data from the 2008–2009 and 2009–2010 seasons (Dataset 7.5), the regression equation is

$$\hat{P} = -14.1 + 1.72S + 0.46F$$

with an $R^2$ value of 40.6%.

Based on this model, we might ask which is more important for team scoring: shooting percentage or free-throw shooting percentage? Moreover, how much more important is the more-important predictor than the less-important predictor? We consider several approaches to answering these questions.

Both predictors are percentages, with similar interpretations; hence, we might compare the effect of increasing shooting percentage by 1 percentage point to the effect of increasing free-throw percentage by 1 percentage point. Using the regression equation, we see that increasing $S$ by 1 unit and holding $F$ constant corresponds to an increase in the average value of $P$ of 1.72 points per game; increasing $F$ by 1 unit and holding $S$ constant corresponds to an increase of 0.46 points per game. Therefore, the effect of a 1-unit increase in shooting percentage is 3.7 times as great as it is for free-throw percentage; that is, $S$ is 3.7 times as important, in some sense, as $F$ in influencing $P$.

However, are these increases of 1 percentage point really equivalent? For the seasons considered, the values of $S$ for the teams ranged from 42.9 to 50.4, and the values of $F$ ranged from 69.2 to 82.4. Therefore, it may be "easier" to increase $F$ than $S$. Using these ranges as a guide, we might conclude that a 1-point increase in $S$ is roughly equivalent to a 1.8-point increase in $F$ because

$$(82.4 - 69.2)/(50.4 - 42.9) = 1.8.$$

An increase in $F$ of 1.8 corresponds to an 0.83-point increase in $P$ on average $(1.8 \times 0.46 = 0.83)$, which can be compared to the 1.72-point increase in $P$ corresponding to an increase in $S$ of 1. Because

$$\frac{1.72}{0.83} = 2.1,$$

based on this analysis we would conclude that $S$ is about 2.1 times as important as $F$.

Alternatively, we could use standard deviation in place of ranges here; the standard deviation of $S$ based on these data is 0.0141, and the standard deviation of $F$ is 0.0261. Suppose we increase $S$ by 1 standard deviation; this corresponds to an increase in $P$ of

$$(0.0141)(1.72) = 0.0243.$$

An increase of 1 standard deviation in $F$ corresponds to an increase in $P$ of

$$(0.0261)(0.46) = 0.0120.$$

It follows that $S$ is about twice as important as $F$, roughly the same conclusion as in the range-based analysis.

Using an analysis based on standard deviations in this way is equivalent to evaluating predictor variable $j$ based on $b_j S_j$, where $b_j$ is the coefficient of variable $j$ in the regression model and $S_j$ is the standard deviation of variable $j$. Such an approach is closely related to *standardized regression coefficients*. The standardized regression coefficient corresponding to $b_j$ is defined as $b_j S_j / S_y$, where $S_y$ is the standard deviation of the response variable in the regression model. It is easy to show that standardized regression coefficients do not change if either the predictor or response variable is multiplied by a constant. Because $S_y$ is the same for each predictor, comparing predictors by comparing standardized regression coefficients is the same as comparing them based on $b_j S_j$.

The preceding analysis illustrates some important general points. One is that comparing predictors based on their coefficients (i.e., the effect of changing the predictor) is sensitive to the scaling used for the variables. The scoring example was particularly simple in this regard because both predictors represent the percentages of shots that are successful. Even so, it is not clear that increasing $S$ by 1 unit should be considered to be equivalent to increasing $F$ by 1. It is important to note that rarely is there one scaling that can be considered to be "correct"; it is up to the analyst to choose the scaling that seems to best reflect the goals of the analysis.

Another important point is that slope-based measures of relative importance rely on the concept of changing one variable while keeping the other variables constant. In the scoring example, such an approach is easy to interpret because shooting percentage and free-throw percentage are approximately uncorrelated ($r = -0.091$). However, if the predictors are highly correlated, it might be unrealistic to consider a hypothetical scenario in which one variable changes while the others remain constant.

Another approach to measuring relative importance is to partition some feature of the response variable among the predictor variables. Let $\bar{Y}$ denote the sample mean of the response variable, and let $\bar{X}_1, \ldots, \bar{X}_p$ denote the sample means of the predictors. It may be shown that the regression line based on the least-squares estimates always passes through the sample means so that

$$\bar{Y} = a + b_1 \bar{X}_1 + \cdots + b_p \bar{X}_p.$$

Therefore, $b_j \bar{X}_j$ represents the contribution of predictor $j$ to the mean of the response. Consider the scoring example: Here, $\bar{P} = 100$, $\bar{S} = 0.460$, and $\bar{F} = 0.765$. It follows that shooting percentage contributes 79.1% to $\bar{P}$, and free-throw percentage contributes 35.2%, so, according to this analysis, shooting percentage is about 2.2 times as important as free-throw percentage. This is in close agreement with the previous analysis. Note that the coefficients of some predictors might be negative, in which case we use the absolute values of the coefficients. This is equivalent to changing the sign of the predictor to change the sign of the coefficient.

Alternatively, we can partition the $R^2$ value for the regression. It can be shown that

$$R^2 = (100 \ b_1 S_1 r_{Y1} + \cdots + 100 \ b_p S_p r_{Yp})/S_Y$$

where $R^2$ denotes the $R^2$ value of the regression, $b_j$ is the regression coefficient of variable $j$, $S_j$ is the standard deviation of variable $j$, $r_{Yj}$ is the correlation of $Y$ and $X_j$, and $S_Y$ is the standard deviation of $Y$. The constant 100 in the equation converts the expression on the right from a proportion to a percentage. It follows that

$$100 \ b_j \ S_j \ r_{Yj}/S_Y$$

is the contribution of variable $j$ to $R^2$.

For example, in the scoring example, $r_{PS} = 0.566$, $S_S = 1.41$, $r_{PF} = 0.239$, $S_F = 2.61$, and $S_Y = 4.08$. Then, $R^2$ can be partitioned as

$$40.6 = [100(1.72)(1.41)(0.566) + 100(0.459)(2.61)(0.239)]/4.08.$$

It follows that the contribution of $S$ to $R^2$ relative to that of $F$ is

$$\frac{(1.72)(1.41)(0.566)}{(0.459)(2.61)(0.239)} = 4.8$$

so that, according to this analysis, shooting percentage is nearly five times more important in scoring than is free-throw percentage.

Although this method of evaluating relative performance is considered by some to be the most theoretically sound approach, there are some features to keep in mind when using it. For instance, it is reasonable to assume that if we remove a variable from the regression equation, then the $R^2$ will decrease by the proportion corresponding to that variable's contribution. However, that is only true if the predictors are uncorrelated with each other.

Another issue is that this approach uses the standard deviations of the predictors, which can be sensitive to the population being studied. For instance, suppose we are modeling the performance of NBA centers in terms of a number of predictor variables, one of which is height. Because there is relatively little variation in the heights of NBA centers, an analysis based on decomposition of $R^2$ might suggest that height is a relatively unimportant predictor of performance. That conclusion is reasonable, provided that the heights in question are typical heights for NBA centers. However, it would not be reasonable to conclude that a very short player could be an effective NBA center.

# 7.14 FACTORS AFFECTING THE SCORES OF PGA GOLFERS

Success on the PGA (Professional Golfers' Association) tour requires expertise in a number of areas, ranging from the power needed for driving to the steady hand needed for effective putting. In this section, we attempt to measure the relative importance of various skills used by golfers on the PGA tour.

The subjects in this example are the 186 golfers for whom the PGA compiled detailed statistics for the 2011 season. The response variable is the player's scoring average $Y$. The predictor variables are average driving distance in yards ($D$), greens in regulation percentage ($G$), average strokes gained in putting ($P$), a statistic designed to measure how many shots are saved per hole by a player's putting ability, and scrambling

**TABLE 7.2** Analysis of golfing data

| VARIABLE | MEAN | SD | CORRELATION WITH SCORING AVERAGE | REGRESSION COEFFICIENT |
|---|---|---|---|---|
| Scoring average | 70.87 | 0.684 | | |
| Driving distance | 291.09 | 8.34 | −0.162 | −0.0252 |
| Greens in regulation | 65.52 | 2.51 | −0.478 | −0.1340 |
| Putting | 0.0341 | 0.356 | −0.399 | −0.6140 |
| Scrambling | 57.57 | 3.34 | −0.568 | −0.1023 |

percentage ($S$), the percentage of holes on which a golfer scores par or better after failing to hit a green in regulation. The data are taken from PGATOUR.com and are available in Dataset 7.6.

The regression equation is

$$\hat{Y} = 92.9 - 0.0252 \, D - 0.134 \, G - 0.614 \, P - 0.102 \, S$$

with an $R^2$ value of 69.6%. Further details on the variables and their relationships are given in Table 7.2. This example presents more difficulties in evaluating the relative importance of the predictors because the units of the variables are very different and difficult to compare directly.

Suppose that each predictor variable is increased by 1 standard deviation. The absolute values of the change in average scoring average are 0.21 for driving distance, 0.34 for greens in regulation, 0.22 for putting, and 0.34 for scrambling percentage. Therefore, according to this analysis, greens in regulation and scrambling percentage are about equally important, and driving distance and putting are about equally important, being two-thirds as important as greens in regulation and scrambling.

Now, consider partitioning the $R^2$ value among the four predictor variables. The $R^2$ value for the regression is 69.6%. The contribution of driving distance is 5.0, of greens in regulation is 23.5, of putting is 12.8, and of scrambling percentage is 28.3; note that the sum of these four values is 69.6. Therefore, scrambling percentage and greens in regulation are the most important predictors of scoring average, with driving distance having relatively little contribution, at least according to this analysis.

Note that the two analyses give the same general conclusions regarding the importance of scrambling percentage and greens in regulation: These are the most important factors in a player's score. Both analyses also conclude that putting is relatively important, although not as important as these two most important factors. The analyses differ regarding driving distance: The analysis based on changing each predictor by one standard deviation concludes that driving distance is roughly as important as putting; the analysis based on partitioning $R^2$, on the other hand, concludes that driving distance is relatively unimportant, being about half as important as putting and about a fifth as important as either scrambling or greens in regulation.

# 7.15 CHOOSING THE PREDICTOR VARIABLES: FINDING A MODEL FOR TEAM SCORING IN THE NFL

Given the vast amount of sports data collected, often there are many potential predictor variables to include in a model. Therefore, an important step in the construction of a regression model is the selection of the predictor variables. This is a well-studied topic in statistics, and many approaches have been proposed; hence, this section contains only a brief summary of some important ideas to keep in mind.

There are two competing considerations when building a regression model: We want the predictor variables to do a good job of explaining the response variable, and we want the model to be as simple as possible. However, even these simple goals are complicated by the fact that, unless the predictor variables are uncorrelated, which is rare with sports data, the meaning of the coefficient of one predictor depends on the other predictors that are in the model.

We illustrate these ideas using the following example: Suppose we are interested in team scoring in the NFL and how it relates to variables measuring offensive performance. Take a team's points per game in a given season as the response variable and measure offensive performance by the variables

- rushing yards per game ($RY$)
- rushing attempts per game ($RA$)
- rushing 20+ yard gains per game ($R20$)
- rushing yards per attempt ($RAVG$)
- passing yards per game ($PY$)
- passing attempts per game ($PA$)
- passing 20+ yards per game ($P20$)
- passing yards per attempt ($PAVG$)
- completion percentage ($COMP$)
- sacks per game ($SCK$)
- first downs per game ($FD$)
- third-down percentage ($THRD$)
- turnovers per game ($TO$)

Defensive takeaways often play an important role in a team's scoring; therefore, it is important to include the variable

- takeaways per game ($TA$)

A regression analysis was performed with points scored per game as the response variable and the 14 predictor variables listed. The data for the analysis are based on the 2009–2011 NFL seasons and are taken from NFL.com; they are available in Dataset 7.7. Choosing the number of seasons to include is an important decision. Using more seasons gives more data values, which results in more accurate estimates. However,

**TABLE 7.3**   Results from a regression analysis using 14 predictors

| PREDICTOR | COEFFICIENT | MARGIN OF ERROR |
|-----------|-------------|-----------------|
| RY | 0.016 | 0.067 |
| RA | 0.013 | 0.236 |
| R20 | 0.054 | 2.277 |
| RAVG | 0.636 | 2.074 |
| PY | 0.049 | 0.135 |
| PA | −0.175 | 0.986 |
| P20 | 0.264 | 1.621 |
| PAVG | 0.587 | 4.904 |
| COMP | −0.066 | 0.221 |
| SCK | −0.472 | 1.300 |
| FD | 0.566 | 0.705 |
| THRD | 0.076 | 0.132 |
| TO | −1.487 | 1.126 |
| TA | 3.710 | 1.24 |

there is the implicit assumption that the relationships among the variables are the same for all the years analyzed, which may not be the case as defenses adapt to new offensive strategies and vice versa. The results of the regression analysis are in Table 7.3.

The $R^2$ value is 85.2%, which is fairly high. However, of the 14 predictors, only turnovers per game and takeaways per game have a coefficient value greater in magnitude than its margin of error. That is, only the coefficients of *TO* and *TA* are statistically significant.

This is most likely because many predictors are highly correlated. For instance, *PY* has correlations 0.82 with *COMP*, 0.81 with *PAVG*, and 0.82 with *P20*. There are at least two important consequences of highly correlated predictors. One is that it is difficult to measure the effect of any one of these predictors; hence, the margins of error of the estimates are inflated. For instance, clearly performance of a team's passing game is closely related to its scoring. However, we have used 5 variables to describe a team's passing game, *PY*, *PA*, *P20*, *PAVG*, and *COMP*. Without vast amounts of data, it is difficult to separate the effects of each of these variables on the response variable. Hence, the individual coefficients are poorly determined; one consequence of this in the present example is that the coefficients of *PY*, *PA*, *P20*, *PAVG*, and *COMP* are all less than half of their margins of error.

A second consequence of highly correlated predictors is that interpretation of the regression coefficients is difficult. For instance, in the example, the coefficient of *COMP* is −0.066, so a higher completion percentage is associated with less scoring. However, it is important to keep in mind that the coefficient applies to increasing completion percentage *holding all other variables constant*. Therefore, to interpret the coefficient of *COMP*, we must think about what it means to increase completion percentage holding passing yards per game, yards per passing attempt, passing plays more than 20 yards, and first downs per game (among others) constant, obviously a difficult task.

**TABLE 7.4**    Regression coefficients for Model II

| PREDICTOR | COEFFICIENT | MARGIN OF ERROR |
|-----------|-------------|-----------------|
| RA | 0.050 | 0.108 |
| RAVG | 1.046 | 0.551 |
| PA | 0.098 | 0.125 |
| PAVG | 2.208 | 0.459 |
| SCK | −0.867 | 0.411 |
| FD | 0.736 | 0.225 |
| TO | −1.537 | 0.539 |
| TA | 3.519 | 0.594 |

It is clear that to obtain a model that is useful for understanding the factors that affect scoring in the NFL we need to eliminate some predictor variables. Although the margins of error of the coefficients provide some useful information for this process, they are best used to supplement careful thought about the nature and properties of the predictors.

Because *PY, PA, P20, PAVG,* and *COMP* all measure aspects of a team's passing game, some of these can be eliminated from the model; I chose to retain *PAVG* and *PA.* Similarly, I eliminated *RY* and *R20* from the model. First downs per game and third-down percentage are also closely related and highly correlated ($r = 0.79$), so I removed *THRD* from the model. Therefore, a regression analysis was performed using only the predictors *RA, RAVG, PA, PAVG, SCK, FD, TO,* and *TA;* call this model, with 8 predictors, Model II and call the original model, with 14 predictors, Model I. The estimated regression coefficients for Model II are in Table 7.4.

The $R^2$ value for Model II is 84.7%; recall that for the original model, the $R^2$ value is 85.2%. In interpreting these values, it is important to realize that, when removing predictors from a model, the $R^2$ value always decreases or, possibly, stays the same; it can never increase. Hence, $R^2$ values are of only limited usefulness in choosing predictor variables for the model.

Several alternatives to $R^2$ have been proposed, designed to "penalize" models that have many predictors. "Adjusted $R^2$," denoted by $R^2_{adj}$, includes an adjustment for the number of predictors in the model; most regression output provides $R^2_{adj}$, making it a convenient choice. However, many analysts prefer the Aikike information criterion (AIC), which can be approximated by

$$R^2_{AIC} = R^2 - \frac{2p}{n}(1 - R^2)$$

where $p$ is the number of terms in the regression model (including the intercept) and $n$ is the sample size. Although choosing a model simply because it has the highest value of $R^2_{adj}$ or $R^2_{AIC}$ is rarely, if ever, a good idea, both of these provide useful information on how well a model fits the data, taking into account the number of predictors in the model.

For Model I, $R^2_{adj} = 82.6\%$ and $R^2_{AIC} = 80.5\%$; for Model II, $R^2_{adj} = 83.2\%$ and $R^2_{AIC} = 81.8\%$. So, by either criterion, the second model fits better than the first. However,

**TABLE 7.5**    Regression coefficients for Model III

| PREDICTOR | COEFFICIENT | MARGIN OF ERROR |
|-----------|-------------|-----------------|
| RAVG | 0.917 | 1.010 |
| PAVG | 2.026 | 0.771 |
| SCK | −0.868 | 0.809 |
| FD | 0.866 | 0.299 |
| TO | −1.467 | 1.041 |
| TA | 3.579 | 1.158 |

further improvement is possible. The two variables that measure attempts, PA and RA, have the only two coefficients that do not exceed their margins of error; moreover, it seems reasonable that attempts do not play an important role in scoring. Hence, PA and RA were eliminated from the model, leading to Model III; the results of a regression analysis using predictors RAVG, PAVG, SCK, FD, TO, and TA are presented in Table 7.5.

For Model III, $R_{adj}^2 = 83.5\%$ and $R_{AIC}^2 = 82.3\%$, both higher than the corresponding values for the first two models. Only the coefficient of RAVG is not statistically significant, although it nearly is. Because a model that does not include a rushing component would be undesirable, removal of RAVG was not considered. Note that significance is greatly affected by sample size, and it is possible that, with more data, the coefficient of RAVG would be statistically significant.

The predictors in this final model are, generally speaking, not highly correlated—the largest correlation (in magnitude) among TA, TO, PAVG, RAVG, and SCK is −0.38 (between TO and PAVG); FD is correlated with RAVG ($r = 0.72$) and SCK ($r = -0.47$). Therefore, TO, PAVG, RAVG, and SCK appear to be measuring different aspects of a team's offense. As would be expected, success in the passing game as reflected in yards gained per passing and sacks per game is closely related to the ability of a team to make first downs, and even after controlling for PAVG and SCK, FD appears to be closely related to scoring.

According to Model III, a takeaway is worth, on average, about 3.6 points, and a turnover is worth about −1.5 points, that is, the team scores about 1.5 points less than it might have otherwise. However, note that a turnover for one team is a takeaway for the opponent. Therefore, any turnover corresponds to, roughly, a 5-point swing (3.6 + 1.5 = 5.1), supporting the general impression of NFL analysts that turnovers are important events in a game. In carrying out such an analysis, it is important to keep in mind the relatively large margins of error for the coefficients of turnovers (about 1) and takeaways (about 1.2). Therefore, although it is clear that turnovers and takeaways are important, a larger sample size would be needed to accurately estimate their effect on scoring; recall that here we are using three seasons of data so that $n = 96$.

Another issue is that, because turnovers and takeaways are relatively rare events, their influence on team scoring is not as large as it might be. For instance, using the approach to measuring relative importance based on partitioning $R^2$, using Model III, the most important variable is FD, accounting for 40.4% of the predictive value of the variables, followed by PAVG with 33.1%; RAVG is the least important with 1.2%. TO

**TABLE 7.6**   Regression results for Model IV

| PREDICTOR | COEFFICIENT | MARGIN OF ERROR |
|-----------|-------------|-----------------|
| RY | 0.072 | 0.023 |
| PY | 0.080 | 0.015 |
| THRD | 0.108 | 0.104 |
| TO | −1.667 | 1.048 |
| TA | 3.731 | 1.129 |

and *TA* are only moderately important, accounting for 13.1% and 7.1% of the predictive value, respectively. Note that, although the values for *TO* and *TA* are relatively small, as noted previously, a turnover for one team is a takeaway for the opponent, so the effect is compounded. Also, it is important to keep in mind that the analysis here is based on season totals, not game-by-game results.

The model reached using these guidelines is not unique. For instance, using the same basic approach but making different decisions at each stage leads to a model with predictors *RY, PY, THRD, TO*, and *TA* (see Table 7.6). This alternative model (Model IV) replaces the yards-per-attempt variables of Model III with total yardage variables and replaces first downs per game by third-down percentage; sacks per game does not appear in the alternative model, but sack yardage is reflected in passing yards per game. Model IV has $R_{adj}^2 = 83.2\%$ and $R_{AIC}^2 = 82.1\%$, very close to the values 83.5% and 82.3%, respectively, for Model III. Note that the coefficients of *TA* and *TO*, which appear in both models, are essentially unchanged. Both models are useful descriptions of the factors affecting scoring in the NFL.

# 7.16  USING REGRESSION MODELS FOR ADJUSTMENT

The regression equation relating the average value of the response variable to the values of the predictor variables can be used to estimate the value of the response variable if it were observed under different conditions than those actually obtained. Therefore, a regression model can be used to construct adjusted statistics that account for variables thought to be important.

Consider the following example: For pitchers, the run average is the average number of runs allowed by the pitcher per 9 innings pitched. Although a pitcher's run average depends primarily on his skills as a pitcher, it also depends on other factors, such as the fielding ability of his team (as well as some luck). Therefore, it may be useful to adjust a pitcher's run average for a factor that is not under his control, such as team fielding.

Let *R* denote a pitcher's run average and let *X* denote a variable for which we would like to adjust. Assuming that *R* and *X* have an approximate linear relationship, we can use the regression model

$$R = a + bX + \varepsilon$$

to relate $R$ and $X$. The coefficient $b$ represents the change in $R$ corresponding to a 1-unit increase in $X$. Therefore, if $X$ is changed to $X_0$, where $X_0$ is some standard value, the estimated change in $R$ is $b(X_0 - X)$, and an adjusted value of $R$ is given by

$$R + b(X_0 - X).$$

Any statistic can be used for $X$ provided that it measures what we are interested in adjusting for and the linear regression model is valid. For instance, suppose that we want to adjust for batting average on balls in play (BABIP), which might be viewed as a measure of team defense.

The correlation between $R$ and *BABIP* is 0.41; the estimated regression equation, using pitchers with at least 150 innings pitched and data from the 2009–2011 seasons (Dataset 7.8), is

$$R = -0.45 + 15.86 \, BABIP.$$

Although this equation can be used to perform the adjustment, in this example, it is more appropriate to fit the model separately for AL and NL pitchers. The estimated regression equations are

$$R = -0.19 + 15.40 \, BABIP \qquad (AL)$$
$$R = -0.89 + 17.03 \, BABIP \qquad (NL)$$

For the AL, the average value of *BABIP* is 0.284; for the NL, it is 0.288. Therefore, the adjusted values of run average, denoted by $R^*$, are given by

$$R^* = R + 15.40 \, (0.284 - BABIP) \qquad (AL)$$

$$R^* = R + 17.03 \, (0.288 - BABIP) \qquad (NL)$$

where $R$ denotes the pitcher's unadjusted run average and *BABIP* denotes the pitcher's BABIP. Note that since only pitchers with 150 or more innings pitched were used to fit the model, the adjustment is only appropriate for pitchers meeting this requirement.

Table 7.7 contains the AL and NL leaders in adjusted and unadjusted run average for the 2011 season.

A major advantage of the regression approach to adjustment over other types of adjustment is that there is considerable flexibility in the variables used for the adjustment. In particular, we are not restricted to adjusting for just one variable; by using a regression model with several predictors, we can adjust for several variables simultaneously. This is illustrated in the following section.

An implicit assumption in this approach is that the relationship between the observed statistic and this adjustment variable is the same for each unit under consideration. For instance, in the adjusted runs allowed example, we assume that the relationship between runs allowed and BABIP is the same for each pitcher. While this is unlikely to be exactly true—for instance, pitchers who give up many home runs might be expected to give up more runs for a given value of BABIP—it might be a reasonable

**TABLE 7.7**   League leaders in adjusted run average for 2011

| | AMERICAN LEAGUE | | | | NATIONAL LEAGUE | | |
|---|---|---|---|---|---|---|---|
| PLAYER | RA | BABIP | ADJUSTED RA | PLAYER | RA | BABIP | ADJUSTED RA |
| Sabathia | 3.00 | .325 | 2.36 | Halladay | 2.35 | .308 | 2.01 |
| Weaver | 2.41 | .254 | 2.87 | Lee | 2.40 | .295 | 2.28 |
| Masterson | 3.21 | .306 | 2.87 | Bumgarner | 3.21 | .331 | 2.48 |
| Wilson | 2.98 | .291 | 2.87 | Kershaw | 2.28 | .275 | 2.50 |
| Jackson | 3.79 | .342 | 2.89 | Vogelsong | 2.71 | .287 | 2.72 |
| Fister | 2.83 | .277 | 2.94 | Lincecum | 2.74 | .288 | 2.74 |
| McCarthy | 3.32 | .303 | 3.03 | Cueto | 2.31 | .257 | 2.84 |
| Gonzalez | 3.12 | .289 | 3.04 | Garza | 3.32 | .313 | 2.89 |
| Hernandez | 3.47 | .311 | 3.05 | Garcia | 3.56 | .327 | 2.90 |
| Verlander | 2.40 | .238 | 3.11 | Carpenter | 3.45 | .318 | 2.94 |

approximation. If this assumption is thought to be unreasonable, we always have the option of modifying the model to make it more appropriate for the goals of the analysis.

# 7.17  ADJUSTED GOALS-AGAINST AVERAGE FOR NHL

Consider data on National Hockey League (NHL) goalies from the 2008–2009, 2009–2010, and 2010–2011 seasons (Dataset 7.9); only goalies with at least 2000 minutes played in a given season are included. Let $G$ denote the goals-against average (GAA) for a goalie in a particular season. A goalie's GAA is greatly affected by the number of shots he faces and the quality of those shots. Therefore, we consider adjustment of $G$ for shots per minute ($S$) and the proportion of those shots that occur when the goalie's team is short handed ($P$).

The estimated regression equation is

$$G = 0.319 + 3.43S + 3.22P.$$

To calculate the adjusted values of $G$, we need standard values of $S$ and $P$; here, we use the average values for the data described. These are 0.503 for $S$ and 0.168 for $P$, leading to the adjusted value of $G$ given by

$$G^* = G + 3.43\,(0.503 - S) + 3.22\,(0.168 - P).$$

Table 7.8 contains the top 10 goalies in terms of adjusted goals against for the 2010–2011 season.

**TABLE 7.8** Top goalies in adjusted goals against for 2010–2011

| PLAYER | GAA | ADJUSTED GAA |
|---|---|---|
| Tim Thomas | 2.00 | 1.94 |
| Roberto Luongo | 2.11 | 2.15 |
| Pekka Rinne | 2.12 | 2.16 |
| Carey Price | 2.35 | 2.27 |
| Jonas Hiller | 2.56 | 2.34 |
| Ilya Bryzgalov | 2.48 | 2.34 |
| Henrik Lundqvist | 2.28 | 2.37 |
| Antti Niemi | 2.38 | 2.40 |
| Jonathan Quick | 2.24 | 2.42 |
| Marc-Andre Fleury | 2.32 | 2.42 |

Note that, in performing an adjustment of this type, it is important to consider the nature of the adjustment variables relative to the goals of the analysis. For instance, suppose we use the same type of analysis used in this section to obtain an adjusted GAA for goalies, to adjust a team's GAA, to better measure a team's defensive capability. The problem with that approach is that an important part of a team's defense is the ability to reduce the number of shots its opponents take, along with their ability to avoid taking penalties. Therefore, by eliminating the effect of these variables from a team's GAA, we are eliminating part of what we are trying to measure.

Specifically, consider the New Jersey Devils in the 2010–2011 season. Their GAA for the season is 2.52 goals per game. Using the same method of adjustment used for goalies (but using team average values of the adjustment variables instead of goalie averages), the adjusted GAA for the Devils is 2.83 goals per game.

The reason why the Devils' adjusted GAA is so much greater than their unadjusted GAA is that they gave up relatively few shots (the fewest in the league) and they took relatively few penalties (also the fewest in the league). If our goal is to measure a team's defensive ability, there is no reason why we should adjust the Devils GAA upward because the team gave up relatively few shots. On the other hand, if our goal is to measure a goalie's ability, such an adjustment might make sense.

The basic rule in these cases is that we should not adjust for a variable that is affected by what we are trying to measure. For instance, we should not measure the offensive ability of NHL players by points scored adjusted for playing time because playing time is affected by the player's offensive ability.

# 7.18 COMPUTATION

To obtain the coefficients of a regression model with several predictors, we simply include all the relevant columns in the $X$ range in the Regression dialog box. For instance, consider the example with runs scored by a given MLB team in a particular

| | A | B | C | D | E | F | G |
|---|---|---|---|---|---|---|---|
| 1 | Y | H | D | T | HR | W | DP |
| 2 | 855 | 0.2554 | 0.0495 | 0.0051 | 0.0335 | 0.0759 | 0.0216 |
| 3 | 875 | 0.2495 | 0.0549 | 0.0055 | 0.0316 | 0.0901 | 0.0212 |
| 4 | 787 | 0.2472 | 0.0477 | 0.0055 | 0.0271 | 0.0836 | 0.0228 |
| 5 | 730 | 0.2489 | 0.0519 | 0.0065 | 0.0206 | 0.0705 | 0.0193 |
| 6 | 762 | 0.2424 | 0.0493 | 0.0035 | 0.0260 | 0.0868 | 0.0271 |
| 7 | 718 | 0.2331 | 0.0488 | 0.0062 | 0.0170 | 0.0901 | 0.0177 |
| 8 | 867 | 0.2303 | 0.0423 | 0.0052 | 0.0352 | 0.0994 | 0.0232 |
| 9 | 721 | 0.2326 | 0.0451 | 0.0051 | 0.0303 | 0.0787 | 0.0186 |
| 10 | 735 | 0.2277 | 0.0437 | 0.0064 | 0.0260 | 0.0884 | 0.0178 |
| 11 | 615 | 0.2345 | 0.0502 | 0.0046 | 0.0154 | 0.0652 | 0.0180 |
| 12 | 708 | 0.2329 | 0.0443 | 0.0021 | 0.0310 | 0.0734 | 0.0250 |

**FIGURE 7.1**   First several rows of the spreadsheet with the MLB offensive data.

season $Y$ as the response variable and hits $H$, doubles $D$, triples $T$, home runs $HR$, walks $W$, and the number of times a player has grounded into a double play $DP$, all measured per plate appearance, as predictors. This example was considered in Section 7.6. Figure 7.1 contains the first several rows of the spreadsheet of these data; the entire data set extends to row 151.

To determine the coefficients of a regression model with $H$, $D$, $T$, $HR$, $W$, and $DP$ as predictors, we use the Regression procedure with the $X$ range specified to include all relevant columns (see Figure 7.2).

The output from the procedure is the same as that from a regression model with one predictor, except that the coefficient table contains one row for each predictor variable; see the discussion of quadratic regression in Section 6.8.

To model interaction using cross-product terms, as discussed in Section 7.8, a column of cross-product terms is created by multiplying two columns, and the new variable is used in the regression model in the usual way.

To use a categorical variable as a predictor in a regression model, we must convert it to one or more indicator variables. Consider the example in which the runs a pitcher allows is modeled as a function of walks, strikeouts, and league using data from the 2009 MLB season; this example was discussed in Section 7.11. Let $R$ denote runs allowed per 9 innings pitched, $K$ denote strikeouts per 9 innings, and $W$ denote walks per 9 innings; let $LG$ denote the league in which the player pitches, taking values "AL" and "NL." The first several rows of the spreadsheet containing these data are given in Figure 7.3.

To convert $LG$ into an indicator variable that can be used in a regression model, we can use the function $IF$. Specifically, the function

$$IF(C2 = AL, 1, 0)$$

is equal to 1 if $C2 = AL$ and is equal to 0 otherwise. Using this function, we can construct a column of indicator variables, as shown in Figure 7.4.

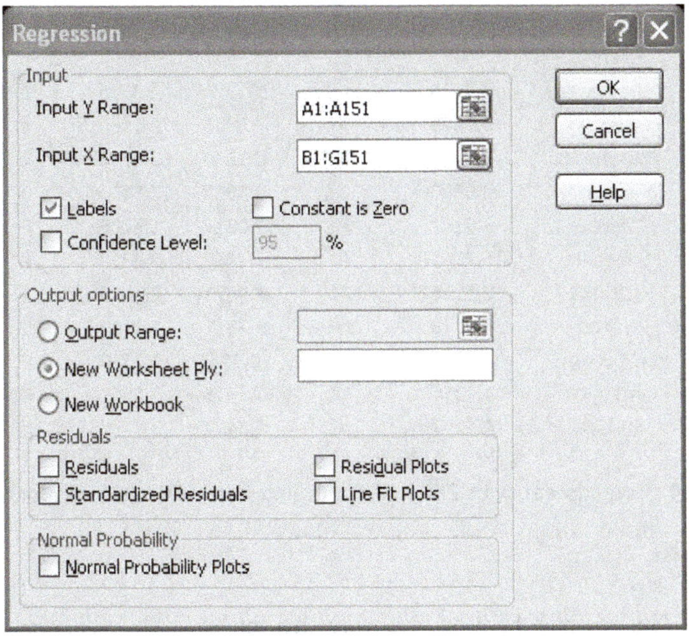

**FIGURE 7.2**   The dialog box for the regression procedure for the MLB example.

| ⊿ | A | B | C | D | E |
|---|---|---|---|---|---|
| 1 | Name | R | LG | K | W |
| 2 | Verlander | 3.7125 | AL | 10.0875 | 2.3625 |
| 3 | Halladay | 3.0879 | AL | 7.8326 | 1.3180 |
| 4 | Hernandez | 3.0545 | AL | 8.1830 | 2.6774 |
| 5 | Wainwright | 2.8970 | NL | 8.1888 | 2.5494 |
| 6 | Lee | 3.4187 | NL | 7.0317 | 1.6705 |
| 7 | Sabathia | 3.7565 | AL | 7.7087 | 2.6217 |
| 8 | Greinke | 2.5116 | AL | 9.4971 | 2.0015 |
| 9 | Haren | 3.2573 | NL | 8.7515 | 1.4913 |
| 10 | Lincecum | 2.7559 | NL | 10.4246 | 2.7160 |
| 11 | Arroyo | 4.1256 | NL | 5.1876 | 2.6551 |
| 12 | Shields | 4.6297 | AL | 6.8422 | 2.1305 |

**FIGURE 7.3**   First several rows of the spreadsheet with the 2009 pitching data.

| ⊿ | A | B | C | D | E | F |
|---|---|---|---|---|---|---|
| 1 | Name | R | LG | K | W | L |
| 2 | Verlander | 3.7125 | AL | 10.0875 | 2.3625 | 1 |
| 3 | Halladay | 3.0879 | AL | 7.8326 | 1.3180 | 1 |
| 4 | Hernandez | 3.0545 | AL | 8.1830 | 2.6774 | 1 |
| 5 | Wainwright | 2.8970 | NL | 8.1888 | 2.5494 | 0 |
| 6 | Lee | 3.4187 | NL | 7.0317 | 1.6705 | 0 |
| 7 | Sabathia | 3.7565 | AL | 7.7087 | 2.6217 | 1 |
| 8 | Greinke | 2.5116 | AL | 9.4971 | 2.0015 | 1 |
| 9 | Haren | 3.2573 | NL | 8.7515 | 1.4913 | 0 |
| 10 | Lincecum | 2.7559 | NL | 10.4246 | 2.7160 | 0 |
| 11 | Arroyo | 4.1256 | NL | 5.1876 | 2.6551 | 0 |
| 12 | Shields | 4.6297 | AL | 6.8422 | 2.1305 | 1 |

**FIGURE 7.4**   Spreadsheet with 2009 MLB pitching data with an indicator variable for league.

These data can now be used in a regression model in the usual way. If the categorical variable takes more than two values, several indicator variables are needed, as discussed in Section 7.11. Each indicator variable can be constructed using the *IF* command used to construct $L$.

# 7.19 SUGGESTIONS FOR FURTHER READING

A detailed treatment of multiple regression models, containing many of the topics discussed in this chapter, is given by Agresti and Finlay (2009, Chapters 10 and 11). See also the work of Snedecor and Cochran (1980, Chapter 17). Mosteller and Tukey (1977, Chapter 13) discuss several important issues regarding regression coefficients in regression models with several predictors.

The use of categorical variables as predictors is discussed by Agresti and Finlay (2009, Section 13.2).

Many different methods of measuring the relative importance of predictor variables have been proposed. A useful general discussion, describing some of the methods presented in Section 7.13, is provided by Achen (1982, Chapter 6). The method based on partitioning $R^2$ is attributed to Pratt (1987); see the work of Thomas, Hughes, and Zumbo (1998) for further discussion of its properties.

The problem of choosing the predictor variables to include in the model is an important one, and like the relative importance of predictors, many solutions have been proposed. Agresti and Finlay (2009, Section 14.1) and Achen (1982, Chapter 5) provide good general discussions that go beyond what is described in Section 7.15. A comprehensive treatment is given in Miller (2002). When using the data to select a statistical model, there is always the danger of "overfitting," in which the model chosen is based on random features of the data rather than on the genuine underlying relationship between the variables; see the work of Freedman (2009, Section 5.8) for further discussion.

The use of regression models to adjust data is discussed by Snedecor and Cochran (1980, Chapter 18).

# References

Achen, C. H. (1982). *Interpreting and Using Regression.* Beverly Hills, CA: Sage.

Agresti, A., and Finlay, B. (2009). *Statistical Methods for the Social Sciences,* 4th ed. Upper Saddle River, NJ: Pearson.

Bishop, Y. M. M., Fienberg, S. E., and Holland, P. W. (1975). *Discrete Multivariate Analysis: Theory and Practice.* Cambridge, MA: MIT Press,

Chatfield, C. (2003). *The Analysis of Time Series: An Introduction,* 6th ed. Boca Raton, FL: CRC Press.

Click, J. (2006). What if Rickey Henderson had Pete Incaviglia's legs? In J. Keri (Ed.), *Baseball between the Numbers: Why Everything You Know about the Game Is Wrong* (pp. 112–126). New York: Basic Books.

Cochran, W. G. (1968). The effectiveness of adjustment by sub-classification in removing bias in observational studies. *Biometrics* **24**, 295–313.

Cox, D. R., and Donnelly, C. A. (2011). *Principles of Applied Statistics.* New York: Cambridge University Press.

Freedman, D. A. (1999). Ecological inference and the ecological fallacy. Retrieved from http://*www.stanford.edu/class/ed260/freedman549.pdf.*

Freedman, D. A. (2009). *Statistical Models: Theory and Practice.* New York: Cambridge University Press.

Gilovich, T. (1991). *How We Know What Isn't So: The Fallibility of Human Reason in Everyday Life.* New York: Free Press.

Goldner, K. (2012). A Markov model of football: using stochastic processes to model a football drive. *Journal of Quantitative Analysis in Sports* **8**, 1–18.

Grinstead, C. M., and Snell, J. L. (1997*). Introduction to Probability,* 2nd rev. ed. Providence, RI: American Mathematical Society.

Huff, D. (1993). *How to Lie with Statistics.* New York: Norton.

Joyner, K. C. (2008). *Blindsided: Why the Left Tackle Is Overrated and Other Contrarian Football Thoughts.* Hoboken, NJ: Wiley.

Keri, J. (Ed.) (2006). *Baseball between the Numbers: Why Everything You Know About the Game Is Wrong.* New York: Basic Books.

Lederer, R. (2009). Using Z-scores to rank pitchers. Retrieved from http://baseballanalysts.com/archives/2009/02/using_zscores_t.php.

Lependorf, D. (2012). Is there a better way of comparing players between historical eras? Retrieved from http://www.hardballtimes.com/main/article/is-there-a-better-way-of-comparing-players-between-historical-eras/.

McClave, J. T., and Sincich, T. (2006). *A First Course in Statistics,* 9th ed. Upper Saddle River, NJ: Pearson.

Miller, A. J. (2002). *Subset Selection in Regression,* 2nd ed. Boca Raton, FL: CRC Press.

Mlodinow, L. (2008). *The Drunkard's Walk: How Randomness Rules Our Lives.* New York: Vintage Books.

Moore, D., and McCabe, G. (2005). *Introduction to the Practice of Statistics,* 4th ed. New York: Freeman.

Moskowitz, T. J., and Wertheim, L. J. (2011). *Scorecasting: The Hidden Influences Behind How Sports Are Played and Games Are Won.* New York: Three Rivers Press.

Mosteller, F., and Tukey, J. W. (1977). *Data Analysis and Regression: A Second Course in Statistics.* Upper Saddle River, NJ: Pearson.

Pratt, J. W. (1987). Dividing the indivisible: Using simple symmetry to partition variance explained. In T. Pukkila and S. Puntanen (Eds.), *Proceedings of the Second International Tampere Conference in Statistics*, pp. 245–250. Department of Mathematical Sciences/Statistics, University of Tampere, Tampere, Finland.

Robinson, W. S. (1950). Ecological correlations and the behavior of individuals. *American Sociological Review* **15**, 351–357.

Rodgers, J. L., and Nicewander, W. A. (1988). Thirteen ways to look at the correlation coefficient. *The American Statistician* **42**, 59–66.

Ross, S. (2006). *A First Course in Probability*, 7th ed. Upper Saddle River, NJ: Pearson.

Schilling, M. F. (1990). The longest run of heads. *The College Mathematics Journal* **21**, 196–207.

Silver, N. (2006a). Batting practice: Is Barry Bonds better than Babe Ruth? In J. Keri (Ed.), *Baseball between the Numbers: Why Everything You Know about the Game Is Wrong* (pp. xxxvii–lxii). New York: Basic Books.

Silver, N. (2006b). Is David Ortiz a clutch hitter? In J. Keri (Ed.), *Baseball between the Numbers: Why Everything You Know about the Game Is Wrong* (pp. 14–34). New York: Basic Books.

Silver, N. (2006c). What do statistics tell us about steroids? In J. Keri (Ed.), *Baseball between the Numbers: Why Everything You Know about the Game Is Wrong* (pp. 326–342). New York: Basic Books.

Silver, N. (2012). *The Signal and the Noise: Why So Many Predictions Fail—But Some Don't*. New York: Penguin Press.

Snedecor, G. W., and Cochran, W. G. (1980). *Statistical Methods*, 7th ed. Ames: Iowa State University Press.

Tango, T. M., Lichtman, M. G., and Dolphin, A. E. (2007). *The Book: Playing the Percentages in Baseball*. Washington, DC: Potomac Books.

Thomas, D. R., Hughes, E., and Zumbo, B. D. (1998). On variable importance in linear regression. *Social Indicators Research* **45**, 253–275.

Tufte, E. R. (2001). *The Visual Display of Quantitative Information*. Cheshire, UK: Graphics Press.

Verducci, T. (2013). Virtue, and victory, no longer synonymous with patience at the plate. Retrieved from http://sportsillustrated.cnn.com/mlb/news/20130423/joey-votto-jayson-werth-taking-pitches/.

Wainer, H. (1989). Eelworms, bullet holes, and Geraldine Ferraro: Some problems with statistical adjustment and some solutions. *Journal of Educational Statistics* **14**, 121–140.

Wardell, P. (2011). 20 statistical oddities from the 2011 MLB season so far. Retrieved from http://bleacherreport.com/articles/687498-20-statistical-oddities-from-the-2011-season-so-far.

Winston, W. L. (2009). *Mathletics: How Gamblers, Managers, and Sports Enthusiasts Use Mathematics in Baseball, Basketball, and Football*. Princeton, NJ: Princeton University Press.

Woolner, K. (2006). Are teams letting their closers go to waste? In J. Keri (Ed.), *Baseball between the Numbers: Why Everything You Know about the Game Is Wrong* (pp. 58–73). New York: Basic Books.

# Available Datasets

The following datasets are available using the link on the website http //www.taseverini.com. Each dataset is in Excel format and is analyzed in the text, as indicated here by the sections listed under "Reference." The data were taken from the website listed under "Source"; however, all of the data are commonly available on many sports-related websites.

2.1   Tom Brady's Passing Yards by Game in the 2001–2011 Seasons
      Column A: Passing yards by game
      Reference: Section 2.3
      Source: Pro-Football-Reference.com

2.2   Jamaal Charles's Rushing Yards by Attempt in the 2010 Season
      Column A: Yards gained by attempt
      Reference: Section 2.3
      Source: Play-by-play data from ESPN.com

2.3   Shooting Percentage and Position of 2010–2011 NBA Players with at Least 300 FG
      Column A: Shooting percentage
      Column B: Position: SF (small forward), PF (power forward), C (center), G (guard)
      Reference: Section 2.3
      Source: ESPN.com

2.4   Batting Statistics for 2011 MLB Players with at Least 502 Plate Appearances
      Column A: Hits
      Column B: Doubles
      Column C: Triples
      Column D: Home runs
      Column E: RBI (runs batted in)
      Column F: Runs
      Column G: Walks
      Column H: Strikeouts
      Column I: Stolen bases
      Column J: Batting average
      Column K: Slugging average
      Reference: Sections 2.4, 5.2
      Source: Baseball-Reference.com

2.5    Winnings of the Top 100 PGA Golfers in 2012
       Column A: Rank
       Column B: Name
       Column C: Winnings in dollars
       Column D: Winnings in log-dollars
       Reference: Section 2.8
       Source: ESPN.com

2.6    Home Runs and At Bats of 2011 MLB Players with at Least 502 Plate
       Appearances
       Column A: At bats
       Column B: Home runs
       Reference: Section 2.9
       Source: Baseball-Reference.com

3.1    Statistics on NFL Field Goal Kickers in the 2011 Season
       Column A: Name
       Column B: Field goals made
       Column C: Field goals attempted
       Column D: Field goals made from 20 to 29 yards
       Column E: Field goals attempted from 20 to 29 yards
       Column F: Field goals made from 30 to 39 yards
       Column G: Field goals attempted from 30 to 39 yards
       Column H: Field goals made from 40 to 49 yards
       Column I: Field goals attempted from 40 to 49 yards
       Column J: Field goals made from 50 yards or greater
       Column K: Field goals attempted from 50 yards or greater
       Reference: Section 3.12
       Source: NFL.com

4.1    Kevin Durant's Game-by-Game Results for the 2011–2012 NBA Season
       Column A: Rebounds
       Column B: Assists
       Column C: Turnovers
       Column D: Personal fouls
       Column E: Points
       Reference: Section 4.3, 4.6
       Source: ESPN.com

4.2    Joe Flacco's Game-by-Game Results for the 2012 NFL Season
       Column A: Completions
       Column B: Pass attempts
       Column C: Passing yards
       Column D: Touchdowns
       Column E: Interceptions
       Reference: Section 4.5
       Source: NFL.com

4.3    LeBron James's Game-by-Game Results for the 2011–2012 NBA Season
        Column A: Rebounds
        Column B: Assists
        Column C: Turnovers
        Column D: Personal fouls
        Column E: Points
        Reference: Section 4.3, 4.6
        Source: ESPN.com

5.1    Rebounding Statistics for 2011–2012 NBA Players with at Least 56 Games
        Played
        Column A: Offensive rebounds per game
        Column B: Defensive rebounds per game
        Reference: Section 5.2
        Source: ESPN.com

5.2    Passing Statistics for 2009 NFL Quarterbacks with at Least 160 Passing
        Attempts
        Column A: Name
        Column B: Attempts
        Column C: Touchdown passes
        Column D: Times sacked
        Reference: Section 5.2
        Source: ESPN.com

5.3    Games Won by MLB Teams in 2010 and 2011
        Column A: Team
        Column B: Wins in 2010
        Column C: Wins in 2011
        Reference: Section 5.2
        Source: ESPN.com

5.4    Points Scored and Points Allowed by 2011 NFL Teams
        Column A: Team
        Column B: Points scored
        Column C: Points allowed
        Reference: Section 5.2
        Source: ESPN.com

5.5    2011 NFL Draft Position and 2011 Wins
        Column A: Team
        Column B: Draft position in the spring, 2011 NFL draft
        Column C: Wins in the 2011 NFL season
        Reference: Section 5.2
        Source: NFL.com

5.6    First- and Second-Half Performances of 2008–2012 MLB Teams
       Column A: Team
       Column B: Year
       Column C: Wins, first half of the season
       Column D: Losses, first half of the season
       Column E: Runs scored, first half of the season
       Column F: Runs allowed, first half of the season
       Column G: Wins, second half of the season
       Column H: Losses, second half of the season
       Column I: Runs scored, second half of the season
       Column J: Runs allowed, second half of the season
       Reference: Section 5.3
       Source: ESPN.com

5.7    Offensive Performance of NFL Teams in the 2011 Season
       Column A: Team
       Column B: Points scored
       Column C: Rushing yards gained by the team's top rusher
       Column D: Passer rating of the team's starting quarterback
       Column E: League ranking of the team's points scored
       Column F: League ranking of the yards gained by the team's top rusher
       Column G: League ranking of the passer rating of the team's starting quarterback
       Reference: Section 5.5
       Source: Pro-Football-Reference.com

5.8    Offensive Performance of NFL Teams in the 1975 Season
       Column A: Team
       Column B: Points scored
       Column C: Rushing yards gained by the team's top rusher
       Column D: Passer rating of the team's starting quarterback
       Column E: League ranking of the team's points scored
       Column F: League ranking of the yards gained by the team's top rusher
       Column G: League ranking of the passer rating of the team's starting quarterback
       Reference: Section 5.5
       Source: Pro-Football-Reference.com

5.9    Pitching Statistics for 2009 MLB Pitchers with at Least 40 Innings Pitched
       Column A: Name
       Column B: Throws: left-handed (L), right-handed (R)
       Column C: Innings pitched
       Column D: ERA (earned run average)
       Column E: Hits allowed
       Column F: Hits allowed per 9 innings
       Column G: Earned runs
       Column H: Home runs allowed
       Column I: Home runs allowed per 9 innings

Column J: Walks allowed
Column K: Walks allowed per 9 innings
Column L: Runners left on base
Column M: Runners left on base per 9 innings
Column N: WHIP (walks plus hits per inning pitched)
Reference: Sections 5.6, 5.7
Source: BaseballGuru.com

5.10  Chris Paul's Game-by-Game Scoring in the 2011–2012 NBA Season
Column A: Game number
Column B: Points scored
Reference: Section 5.8
Source: ESPN.com

5.11  Offensive Statistics for 2009–2012 MLB Teams
Column A: Team
Column B: Year
Column C: RBI (runs batted in) per plate appearance
Column D: Runs scored per plate appearance
Column E: Home runs per plate appearance
Column F: Doubles per plate appearance
Reference: Section 5.13
Source: Baseball-Reference.com

5.12  Offensive Statistics for 2009–2012 MLB Players with at Least 502 Plate
Appearances
Column A: Year
Column B: RBI (runs batted in) per plate appearance
Column C: Runs scored per plate appearance
Column D: Home runs per plate appearance
Column E: Doubles per plate appearance
Reference: Section 5.13
Source: Baseball-Reference.com

5.13  Three-Point Shooting Percentage and Scoring for 2010–2011 to 2012–2013
NBA Teams
Column A: Team
Column B: Year: 2011 (2010–2011), 2012 (2011–2012), 2013 (2012–2013)
Column C: Three-point shooting percentage
Column D: Points per game
Reference: Section 5.13
Source: Basketball-Reference.com

5.14  Three-Point Shooting Percentage and Scoring for 2010–2011 to 2012–2013
Qualifying NBA Players
Column A: Team

Column B: Year: 2011 (2010–2011), 2012 (2011–2012), 2013 (2012–2013)
Column C: Three-point shooting percentage
Column D: Points per game
Note: Qualifying players must have either 56 games played or 1126 points scored
Reference: Section 5.13
Source: ESPN.com

5.15    Pitches per Plate Appearance and OBP (On-Base Percentage) for 2010–2012
MLB Teams
Column A: Team
Column B: Year
Column C: OBP
Column D: Pitches per plate appearance
Reference: Section 5.14
Source: Baseball-Reference.com

5.16    Pitches per Plate Appearance and OBP (On-Base Percentage) for 2010–2012
MLB Players with at Least 502 Plate Appearances
Column A: Player
Column B: Year
Column C: OBP
Column D: Pitches per plate appearance
Reference: Section 5.14
Source: Baseball-Reference.com

6.1     Offensive Statistics for 2007–2011 MLB Teams
Column A: Team
Column B: Year
Column C: Runs
Column D: Triples
Column E: Home runs
Column F: Stolen bases
Column G: OPS (on base plus slugging)
Reference: Section 6.2, 6.3
Source: MLB.com

6.2     Rebounds per Game and Height of 2010–2011 NBA Centers with at Least 70
Games Played
Column A: Player
Column B: Rebounds per game
Column C: Height in inches
Reference: Section 6.3
Source: Basketball-Reference.com

6.3   WHIP (Walks Plus Hits per Inning Pitched) and Age for 2011 MLB Pitchers
      with at Least 36 Innings Pitched
      Column A: Age in years on June 30, 2011
      Column B: WHIP
      Reference: Section 6.3
      Source: BaseballGuru.com

6.4   Ken Griffey Jr.'s Home Run Rate by Year
      Column A: Year
      Column B: Home runs per 100 at bats
      Column C: Year of his career
      Reference: Section 6.8
      Source: Baseball-Reference.com

6.5   Points and Minutes Played for 2011–2012 NHL Forwards with at Least 60
      Games Played and 6 Minutes Played per Game
      Column A: Name
      Column B: Games played
      Column C: Points (goals plus assists)
      Column D: Time on ice per game, in minutes
      Reference: Sections 6.8, 6.10
      Source: NHL.com

6.6   Games Won and Lost and Runs Scored and Allowed for 2008–2012 MLB
      Teams
      Column A: Team
      Column B: Year
      Column C: Wins
      Column D: Losses
      Column E: Runs scored
      Column F: Runs allowed
      Reference: Section 6.10
      Source: ESPN.com

7.1   Performance of 2009 MLB Pitchers with at Least 40 Innings Pitched
      Column A: Strikeouts per 9 innings pitched
      Column B: Walks per 9 innings pitched
      Column C: League: AL (1), NL (0)
      Column D: Home runs allowed per 9 innings pitched
      Column E: WHIP (walks plus hits per inning pitched)
      Column F: Runs allowed per 9 innings pitched
      Reference: Sections 7.3, 7.8, 7.11
      Source: BaseballGuru.com

7.2    Points Scored and Time of Possession for 2009–2011 NFL Teams
       Column A: Team
       Column B: Year
       Column C: Points scored per game
       Column D: Time of possession per game in minutes
       Column E: Yards gained per game
       Reference: Section 7.3
       Source: NFL.com

7.3    Offensive Performance of 2007–2011 MLB Teams
       Column A: Team
       Column B: Year
       Column C: Runs scored
       Column D: Hits per plate appearance
       Column E: Doubles per plate appearance
       Column F: Triples per plate appearance
       Column G; Home runs per plate appearance
       Column H: Walks per plate appearance
       Column I: Double plays per plate appearance
       Reference: Section 7.6
       Source: MLB.com

7.4    Offensive Rebounds and Scoring of 2010–2011 NBA Players with at Least 70
       Games Played
       Column A: Player
       Column B: Points per game
       Column C: Offensive rebounds per game
       Column D: Position: center (C), forward (F), guard (G)
       Reference: Section 7.12
       Source: ESPN.com

7.5    Shooting Percentages and Scoring of 2008–2009 to 2010–2011 NBA Teams
       Column A: Team
       Column B: Year
       Column C: Points per game
       Column D: Field goal percentage
       Column E: Free throw percentage
       Column F: Three-point shooting percentage
       Reference: Section 7.13
       Source: ESPN.com

7.6    Performance of PGA Golfers in the 2011 Season
       Column A: Golfer
       Column B: Average driving distance in yards
       Column C: Percentage of greens in regulation
       Column D: Average strokes gained in putting (see text)

Column E: Scrambling percentage (see text)
Column F: Scoring average
Reference: Section 7.14
Source: PGA.com

7.7    Offensive Performance of NFL Teams for the 2009–2011 Seasons
       Column A: Team
       Column B: Year
       Column C: Points scored per game
       Column D: Takeaways per game
       Column E: Passing yards per attempt
       Column F: Turnovers per game
       Column G: Sacks allowed per game
       Column H: Rushing yards per attempt
       Column I: First downs per game
       Column J: Passing plays of 20 or more yards per game
       Column K: Running plays of 20 or more yards per game
       Column L: Completion percentage
       Column M: Third-down success percentage
       Column N: Passing attempts per game
       Column O: Rushing attempts per game
       Column P: Rushing yards per game
       Column Q: Passing yards per game
       Reference: Section 7.15
       Source: NFL.com

7.8    Run Average and BABIP (Batting Average on Balls In Play) for 2009–2011
       MLB Pitchers with at Least 150 Innings Pitched in a Season
       Column A: Player
       Column B: Year
       Column C: League
       Column D: Runs allowed per 9 innings
       Column E: BABIP
       Reference: Section 7.16
       Source: BaseballGuru.com

7.9    Performance of NHL Goalies from 2008–2009 to 2010–2011 with 2000
       Minutes Played in a Season
       Column A: Player
       Column B: Year
       Column C: Shots faced per minute
       Column D: Proportion of shots while short-handed
       Column E: Goals allowed per 60 minutes
       Reference: Section 7.17
       Source: NHL.com

# Index